FISHING
WITH JOHN

Books by Edith Iglauer

THE NEW PEOPLE: *The Eskimo's Journey into Our Time* (1966)
DENISON'S ICE ROAD (1975)
INUIT JOURNEY (1979)
SEVEN STONES: *A Portrait of Arthur Erickson, Architect* (1981)
THE STRANGERS NEXT DOOR (1991)

EDITH IGLAUER

FISHING *WITH JOHN*

Harbour Publishing

First paperback edition 1992.

Second paperback printing, July 1992

Cover design by Roger Handling, adapted from the design of the hardcover edition by Tere Lo Prete. The chart on the front cover was John Daly's and shows his notations. The inset photograph shows the *Muriel D*, one of the trollers which Daly owned prior to building the *MoreKelp*.

Harbour Publishing
Box 219
Madeira Park, BC V0N 2H0

Parts of this book first appeared, in slightly different form, in *The New Yorker*.

Printed and bound in Canada.

CANADIAN CATALOGUING IN PUBLICATION DATA

Iglauer, Edith.
Fishing with John

ISBN 1-55017-048-1

1. Iglauer, Edith. 2. Daly, John, 1912-1978.
3. Pacific salmon fishing – British Columbia – Anecdotes.
I. Title
HD8039.F66C364 1988 639'.2755'092 C88-91349-5

For John

Acknowledgments

Commercial fishermen and others who earn their living on the water are a fussy lot, so when I began writing I nervously asked three of those whom I knew best to read along as I wrote: to check and cross-check the technical areas of boats and commercial fishing in which I felt my knowledge to be a bit hazy. I soon discovered that the women in their houses were reading over their shoulders, with very sharp eyes. I am extremely grateful to Anne Clemence and Sam Lamont, Ray and Doris Phillips, and Peter and Lorraine Spencer, for going over the manuscript not once but several times; and to Reg Payne, who made a special trip to Garden Bay to examine it when I was finished. Among these regular readers, I must single out Richard Daly as being wonderfully helpful in every way, vivid and precise in his memories of the years he spent fishing with his father.

A special word of appreciation is owed to other fishing experts and friends who gave me their patient attention: Jacqueline Bernstein, Renee Brown, Ruth Chambers, Don Cruickshank, Sean Daly, the late Hubert Evans, Harold Fallon, Jane Iglauer Fallon, Frances Fleming, Anna Hamburger, Philip Hamburger, my sons Jay and Richard Hamburger, David Hardie, Diana Knowles, Emily Maxwell, William Maxwell, Geoff Meggs, David O'Connor, Cecil Reid, Jr., the late Maximiliano Southwell, the late Roger Stanier, and Howard White.

Halfway through the writing of my book, I bought a word processor. In fact, I ended up buying two word processors, one for each coast, since I live a third of the year in the East. I don't know how I wrote books before I had a computer, but, given my shaky mechanical skills, I could not have finished this one on either coast without the expert computer advice of Stephen Osborne in the West or Cathy Blaser in the East.

Author's Note

I have worked with William Shawn twenty-five years, writing for *The New Yorker*. He has told me many times how much he detests cold weather, snow, getting wet, and, in general, fishing. Yet he sent me to the Canadian Arctic many times, and when I proposed to write about my life at sea no one could have been more encouraging. His face has a way of lighting up when he likes an idea, and as the writing unfolds he displays an excitement that, together with his unflagging support, is what keeps a writer going. Part of his genius is his ability to make you see for yourself what you want to say and to hold you to that vision. I am indeed fortunate to have him as an editor and a friend.

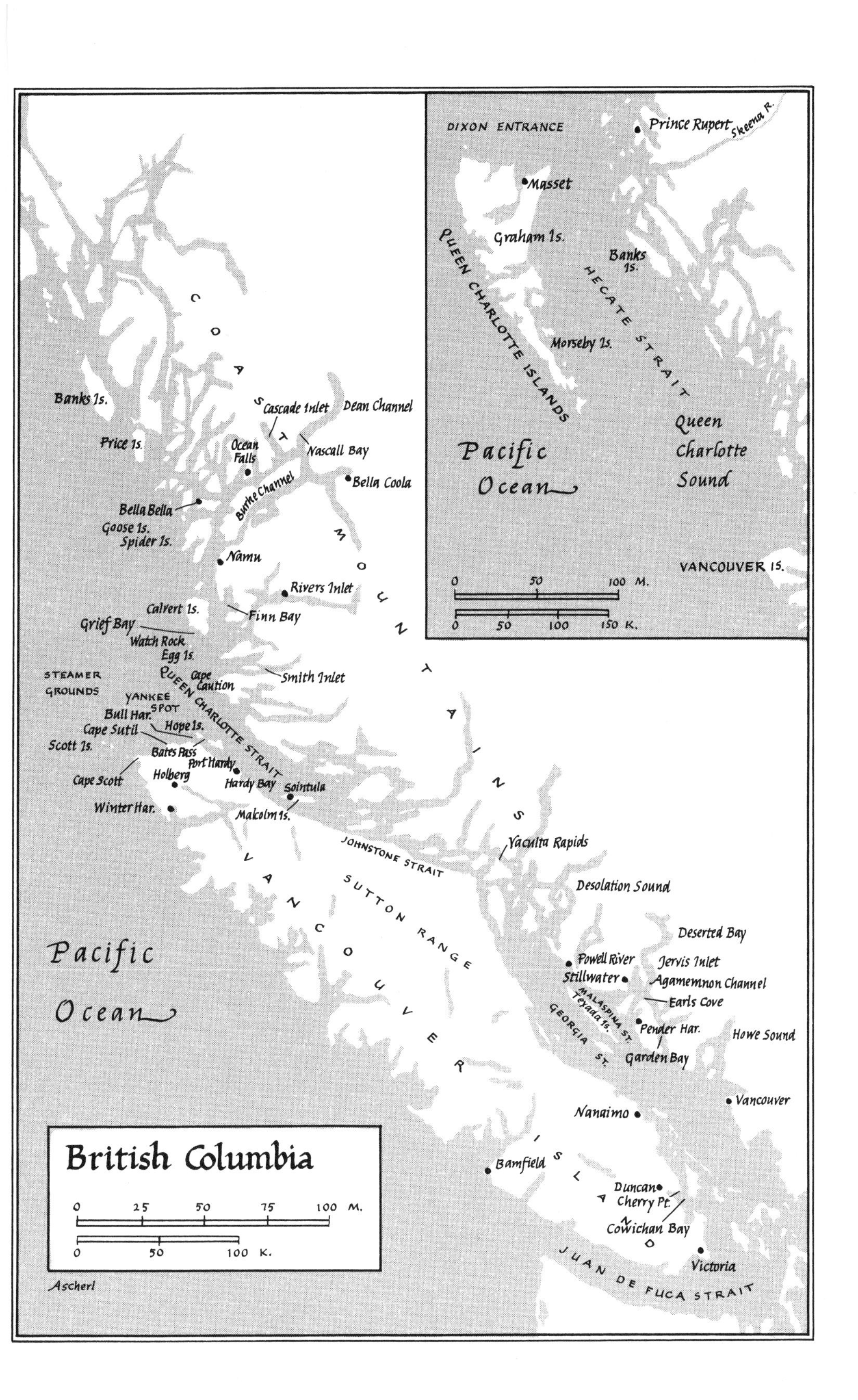
British Columbia
0 25 50 75 100 M.
0 50 100 K.
Ascherl
DIXON ENTRANCE
Prince Rupert
Skeena R.
Masset
Graham Is.
Banks Is.
QUEEN CHARLOTTE ISLANDS
HECATE STRAIT
Morseby Is.
Pacific Ocean
Queen Charlotte Sound
VANCOUVER IS.
0 50 100 M.
0 50 100 150 K.
COAST MOUNTAINS
Banks Is.
Price Is.
Cascade Inlet
Dean Channel
Ocean Falls
Nascall Bay
Bella Coola
Burke Channel
Bella Bella
Goose Is.
Spider Is.
Namu
Rivers Inlet
Calvert Is.
Finn Bay
Grief Bay
Watch Rock
Egg Is.
STEAMER GROUNDS
Cape Caution
Smith Inlet
YANKEE SPOT
Bull Har.
Cape Sutil
Hope Is.
Scott Is.
Bates Pass
QUEEN CHARLOTTE STRAIT
Port Hardy
Cape Scott
Holberg
Hardy Bay
Sointula
Winter Har.
Malcolm Is.
JOHNSTONE STRAIT
Yaculta Rapids
Desolation Sound
SUTTON RANGE
VANCOUVER ISLAND
Deserted Bay
Powell River
Jervis Inlet
Stillwater
Agamemnon Channel
MALASPINA ST.
Earls Cove
Texada Is.
GEORGIA ST.
Pender Har.
Howe Sound
Garden Bay
Pacific Ocean
Vancouver
Nanaimo
Bamfield
Duncan
Cherry Pt.
Cowichan Bay
Victoria
JUAN DE FUCA STRAIT

FISHING
WITH JOHN

Introduction

Each year, from spring through fall, a number of small vessels with tall poles stretched out on either side appear, like large birds, on the coastal waters of British Columbia. They seem to sit motionless on the surface, but they are moving gently at a speed of around two knots. They are trollers—with a lacework of lines and hooks hanging into the sea from their poles—searching for salmon. For the last four years of his life, I went fishing with my husband, John Heywood Daly, a commercial salmon fisherman, on his forty-one-foot troller, the *MoreKelp*, registered out of Vancouver, B.C. He fished, which he had been doing steadily for forty years, and I wrote, working part-time as his deckhand.

Writing had been my profession for as many years as John fished, so, after the first strangeness of living on a boat wore off, I started working on a manuscript I had brought along. At the same time, the remarkable turn my life had taken enthralled me, and I began taking notes on what was happening around me and of what we said to one another. On the surface, it looked as if we had very little in common, except that we loved each other. I was born and raised in comfortable circumstances, in a large Midwest American city—Cleveland, Ohio—by loving parents who gave my sister and me every advantage. In my first marriage, I lived an urban, sophisticated life in New York, the wife of a distinguished and very funny writer. Before and after my two sons were born, I worked as a journalist. My rampant curiosity led me to tackle a broad array of subjects. I was fascinated by my assignments, some of which resulted in books. During the Second World War, I covered Eleanor Roosevelt's press conferences for the U.S. Office of War Information and was a newspaper correspondent in Europe. Later, I wrote about the United Nations, air pollution, the New York City mounted police, a Canadian prime minister, a great international architect, and, on many trips to the Northwest Territories of Canada, about Inuit communities in the eastern Arctic and in the West, about an ice road to a silver mine in the high Arctic. It was a privileged existence, replete with friends, art, music, theatre, comfortable living, and fancy restaurants, until I came to British Columbia, and turned my life upside down when I met and married John Daly.

Time on the boat was a continuous surprise: a discovery of John, of facets of myself I had never suspected were there, and of a way of living I could not have imagined before I came on the *MoreKelp*. Nothing was familiar, and I was perpetually astonished by the complexity of a fisherman's life and the skills he had to have for survival at sea. John had to be an expert navigator, meteorologist, electrician, shipwright, and mechanic; and, like most other fishermen, he was an excellent cook. I could also see that John's world was changing. At sixty-one, he had fished alone most of his life, and he was having to adjust to the constant presence of another human being—one who knew nothing about fishing or boats. I had everything to learn.

The notion that I might write about fishing developed because John needed an official reason to have me with him on the *MoreKelp*. By law, anyone on a Canadian commercial fishing vessel during the commercial fishing season had to be a Canadian citizen or have stated the intention of becoming one by taking out "landed immigrant" papers. Any exception required the special permission of the Federal Minister of Fisheries. The first two seasons with John, I obtained permission to be on the *MoreKelp* so that I could write about fishing. During my third year in Canada, I "landed."

One day when John was fishing, perched on the stern gunwale, his long legs resting against the fish box, one hand gripping a heavy rope tied to the boom to steady himself in the rolling sea, I joined him on deck for his second cup of coffee. With his gray hair, gray eyes, glasses, long face, strong aquiline nose, neat white mustache, and close-trimmed beard, he looked scholarly, professorial. In his house, books were everywhere; he had to sweep them aside to make a place to eat.

"I wonder what you would do if you had your life to live over again," I said.

His eyes had been riveted on the springs at the tops of his trolling poles, watching for the nervous shiver that signals a fish striking the hooks below the water. He turned to stare at me in amazement. "I wouldn't do anything different! Certainly not!" he said. "We trollers are the true water gypsies. I have my mountains around me, the water, this boat. I have the sun, the wind, the stars, and the marvelous changes brought about by this northern climate's rain and fog. What will probably get me out of fishing is this damned cold, which I can't deal with as well as I did when I was younger. Meanwhile, could *anything* be better?"

He leaned down to turn the steering wheel, one of four wheels situated at different places on the boat so that he could maneuver it wherever he happened to be. He avoided a floating log, sped up to pass a clump of seaweed, and slowed down again. "You can teach a fisherman navigation, electronics, diesel mechanics, and marine biology, but there's an indefinable factor nobody can learn," he said. "I sometimes feel I am more fish than man—especially when I've been fishing for five or six weeks, with only a few hours off. I become more observant, develop a sixth sense of where to go at the right time to where the fish are."

He turned the boat away from another floating log, which had two passenger seagulls aboard. "I wake up at three in the morning and my whole nervous system wants to go fishing. That desire must go back thousands of years. I believe that some of us who love fishing and have spent our lives at it reach back into the sources of our evolution, which were from the ocean, to tune ourselves in on instincts we possessed millions of years ago. To be a good fisherman, you have to be able to think like a fish."

I

The *MoreKelp* was on dry land the first time I saw her. I had moved from New York City to Vancouver in the early fall, and shortly afterwards I met John Daly through a close mutual friend, Dallas Read. During the Second World War, Dallas had married an Englishman in the Canadian Air Force who became a commercial salmon fisherman off the coast of British Columbia. The most durable result of that brief experiment in marriage and fishing was Dallas Read's friendship with John, which began when their trolling vessels were tied up next to each other at a fish camp up the coast. Dallas subsequently returned to the United States and had not seen John since, although they continued to correspond about events in their lives, which included divorce for both of them, and a second marriage for Dallas. She asked me to call him when I got to Vancouver, which I did.

On his next trip to Vancouver from Pender Harbour, where he lived, on the Sechelt Peninsula, farther north, John came to see me at the apartment I had rented—one that fulfilled my most basic requirement: a view of water and, if possible, mountains, both of which it had in abundance. From then on, we met whenever he came into town, until one day I received a phone call from John inviting me for the first time to visit him at his home—later mine as well—in Pender Harbour. We agreed that I would take the bus the next morning, but five minutes later he called me back. "Take the evening bus tonight," he said. "I want you to see the beautiful sunset." I hesitated, trying to make up my mind, and he shouted into the phone, "We're giving too much business to that bloody monopoly, B.C. Tel! The bus leaves at six-twelve, and that's an order!" Whenever I was indecisive after that, a "six-twelve order" became the catch phrase to straighten me out.

I had never thought to ask him how long the trip to his house took, but since he had mentioned seeing the sunset, I figured I would soon be there. The hours passed, night fell, the bus rolled onto a big boat that carried us over a large body of water, then off onto an endless road. At stops along stretches of unlit highway, passengers departed, until finally the bus contained only me, the driver, and two drunks on their way to a place I gathered from their mumbled conversation was called Powell River. From Vancouver to Pender Harbour is actually a three-hour trip northwest up the Pacific coast, partly by ferry through Howe Sound, an archipelago with a magnificent view, on clear days, of the towering Coast Range, whose mountains rise precipitously from the water's edge—mountains that at night become a vaguely menacing border between sea and sky.

Inside the narrow entrance to Pender Harbour are twelve miles of completely protected coastline, with many bays, a sprinkling of islands and islets, several little settlements with stores and marinas, and innumerable houses down by the water or half hidden among the trees along the rocky cliffs. Some of the names of interior areas are reminders of an adventurous past: Gunboat Bay, Whiskey Slough, and, to the south, Secret Cove, Smuggler Cove, and Buccaneer Bay. Pender Harbour still contains a healthy residue of the loggers and fishermen who originally settled in the communities along its sheltered shores, which are dotted with small "marine ways"—the colloquial expression for the timber framework on which commercial and sportfishermen can have their boats pulled up for repairs and overhauling.

I saw a sign beside the highway that said we were entering the Pender Harbour area, and, not too long after, the bus pulled up at a gas station by a corner crossroad. John was waiting in his green Volvo station wagon when I got off. We drove another eight miles, along a moonlit road that twisted and turned around the shores of a large lake. It skirted dangerously close to edges where, John casually remarked, several cars had gone into the water on icy winter nights. He added, "When you drive around the lake, it might be wise to unfasten your seat belt." It took me six months after I moved to Pender Harbour before I cranked up enough courage to drive his car alone to the main highway past that lake.

John turned onto a shorter road and then in at a driveway marked by two long poles shaped into a V, with a sign hanging in the space at the top which said, "*J Daly,*" and had a painted gray-and-white wooden fish suspended below it. Halfway down a steep hill, he stopped the car beside a small cottage sheathed in rough-hewn Douglas-fir planking, with a cedar shake roof. The cottage was perched on the side of the rocky slope, overlooking the water. We were at John's house.

I got out of the car, walking by a garden patch that contained a blue tin bathtub filled with plantings, and along a strip of rough concrete that ended at wooden steps to the house. The first thing that caught my eye before I climbed them was a weather-beaten board hanging from the front veranda, in which had been carved, in deeply incised letters six inches high:

FAITH IN HUMANITY, NOT GODS OR $

At the top of the steps, John dropped my bag and guided me around to a deck outside, facing the water.

"Where are we?" I asked.

"We're in a corner of Pender Harbour called Garden Bay," he said.

It was a starry night, with a full moon that created a glistening, rippled carpet of black water below. "Whatever happened to that sunset you wanted me to see?" I asked.

John looked puzzled for a moment, then clapped his hand to his head and laughed. "Once I had made up my mind to clean the house, I felt free to invite you here, and I guess then I wanted you to come as soon as possible," he said, hurrying me through the door into the kitchen. "You'll have your sunset now!"

While I wandered through the rooms, from the large cupboard-lined kitchen, with its black-and-white oil stove, to the long, narrow living room paneled with brown airplane spruce, and into a tiny hall with a small windowed bedroom at each end and a hideous bottle-green bathroom in between, John put on the electric kettle to boil water for tea. I heard him scurrying around, and when I came back in the kitchen he had propped a projector on two books on the table and put up a folding screen on a tripod beside the oil stove. He handed me a cup of tea, turned out the lights, and for two hours gave me a slide show of spectacular sunsets he had photographed at sea, reliving each sunset on the screen as if he were seeing it for the first time. I was astonished that I could look for two solid hours at nothing but sunsets, but he was so joyful in his descriptions of the place and circumstance of each shot ("This orange sunset was my reward after enduring a helluva gale that blew for three days and three nights in thick fog off Cape Scott") that it was impossible not to be swept up in his pleasure. "I love being alone because I never am alone," he said while he was putting the slides away. "I am physically alone, yes; but I am together with everything I love all the time. I've had a long love affair with my environment and with fishing. I am at one with the sea and the mountains. By now, I am part of the B.C. coast."

In the morning, John took me in the Volvo on a rapid tour of the area. It included a twenty-minute drive farther north up the main highway, ending at a place called Earls Cove, where a family of fishermen named Earl had lived. The primary feature now of this scenic spot, nestled among fjords and high mountains in the distance, was a quaint car-and-passenger ferry that arrived when we did. We watched a half dozen or so waiting cars drive onto it, and, after the ferry left, John traced on a map in his car the ferry's journey through a winding water path. "It connects the north end of our Sechelt Peninsula with the mainland above—primarily with the town of Powell River and its huge pulp mill," he said. "Using this ferry and the one you came across on last night, it is possible to drive straight through from Vancouver to Powell River."

Right after lunch, having fulfilled his sightseeing duties as host in the morning, John drove me to another place by the shore, five minutes away, where he went back to the work that my arrival had interrupted. He was engaged in his annual chore of scraping and copper-painting the wooden hull of his fishing boat to protect her from teredo worms. The *MoreKelp* had been moved temporarily from her home mooring, an old scow chained to a ramp in the bay below John's house, to a repair dock two bays away. The boat rested in the marine ways' superstructure, a wooden cradle that had been lowered into the water into which the vessel had been maneuvered, the whole then drawn out of the water and up on shore by winch high enough for John, who was six feet four, to scrape and paint underneath. The brass propeller and wooden rudder were plainly visible; indeed, the entire hull—narrow, deep, V-bottomed, pointed bow and squared at the stern with rounded edges—was exposed. It seemed naked and vulnerable without its normal sea cover.

John said, "Go on board and take a look," and immediately vanished beneath the boat, where I could hear him scraping. I walked along the dock beside the marine ways. Above the rusty color of the copper-painted bottom, the sides of the hull were white, with slate-blue on the wide gunwales. A few feet from the bow, the white pilothouse, with its green roof, occupied about a third of the deck space, and just behind it was the mast, with its long boom to the stern. Underneath the boom was a blue lidded box about four feet square. The rest of the deck contained wheel-like contraptions, several lead balls, and other maritime equipment equally strange to me: I had spent a whole lifetime on land. On the outside wood of the stern was a white oblong patch with the words in large black letters:

MOREKELP
VAN. BC

Nothing John had said about his boat had prepared me for this first sight of her. She looked so tiny—scarcely bigger than a large rowboat. Where was the grand ship I had imagined? Was it for the sake of this shabby little vessel that John hovered around radios when he came to Vancouver, listening to every weather report, dashing for the ferry to get home and make sure his boat was tied up securely, at the hint of an approaching storm? How could a man fish miles out on the Pacific Ocean for almost half of each year alone, as he apparently often did, on so small, so fragile a craft? Especially at his age, and after a severe heart attack. It had occurred ten years earlier in the middle of a heavy fishing season, as he delivered fish to the cannery in Prince Rupert, just south of the Alaskan border. He had been hospitalized there for six weeks. Since then, he told me, he no longer fished that far north.

I could see the legs of John's heavy brown wool fish pants on the other side of the hull and hear the scrunch scrunch of his paint scraper. I walked over to the gangplank that was precariously balanced against the crude frame of the marine ways, hesitantly climbed it, teetered on a narrow board at the top, and stepped over the low gunwale onto the deck of the *MoreKelp*. Gripping the smooth wood of the mast, I listened to the reassuring noise of John's presence just below. I noticed black lettering in a firm hand over the wooden doorframe of the pilothouse and I moved closer to read it: LORD, LORD, WASN'T THAT A FISH! PRICE OF EVERYTHING, VALUE OF NOTHING. O.W.

I started to walk through the open doorway of the pilothouse and bumped my head on the low hatch cover shielding the doorframe as I made the one step down from the deck to enter. I went in, rubbing my forehead. Bottles of nuts, bolts, and fishhooks hung from the ceiling, their lids nailed to the pilothouse beams; open shelves were piled with tools, and a large, heavy black vise was attached to the back wall; radios of assorted sizes were suspended from side walls; a pair of old-fashioned long binoculars with the initials "J D" crudely lettered in white paint on either section rested beside the binnacle, the wooden case around the compass on the shelf at the helm; and a cascade of fishlines with rainbow-colored plastic lures and black hooks dripped from the port wall underneath a .22 rifle and a fishing rod, horizontal in a wall bracket. I was standing in the narrow aisle between a scarred wooden cupboard, a miniature sink with a hand pump, and a rusty stove that together occupied most of the floor length to my right; and a long, narrow bunk with a canvas cover on my left. Despite these household familiars, I felt as if I were inside a garage. The smell of diesel oil was pervasive. I have thought a lot since about my first real look at the interior of the *MoreKelp*. For curtains there were cracked green shades on the two small slanted side windows, starboard and port. Comfortable chairs? There was one hard wooden seat to the left of the steering wheel, and a backless stool on the right was folded up to a hook on the wall. Was I looking for charm? Among the hundreds of fishing vessels I have seen since that day, not one has had that elusive interior distinction, although many—most—have an exterior grace in the water. They are workboats, loaded with gear and expensive electronic equipment, designed for only one purpose: to catch fish. The *MoreKelp* was simply the first troller I saw, and I came to understand and love her, with all her miseries and peculiar discomfort, more than any house I have ever had. On the rare occasion when I was left behind to watch from shore as she departed for the fishing grounds, I would be so enchanted with the gentle beauty of the scene she created, gliding through the water as the vertical trolling poles slowly descended to a forty-five-degree angle, unfolding like wings, that I would watch and watch until she was a speck—until she dissolved somewhere in the northern horizon.

On that earliest visit, I sat down on the bunk, feeling uncertain, disconnected; it was so unlike anything I had known. The only sound was the scrape of the blade John was using to remove old paint. Gradually, I perceived that the white walls inside the pilothouse were covered with words printed in bold black letters about an inch or so high: above and below the five large windows toward the bow, even on the inside edges of their sills; above the stove and sink, around the radios, the bunk, and the doorway. Were the walls talking? WATCH FOR A GIGANTIC HOAX, I read by the pilothouse door; then, LAWYERS SPEND THEIR PROFESSIONAL CAREERS SHOVELING SMOKE. O. W. HOLMES; and, under that, IF U GIVE A MAN A FISH HE HAS 1 MEAL, IF U TEACH HIM HOW TO FISH HE CAN FEED HIMSELF 4 LIFE. The space around the front windows of the pilothouse was especially rich in black print. Under the far port window: IDEAS TAKE ON HANDS AND FEET; and between that window and the next one, WE HAVE NO MORE RIGHT TO CONSUME HAPPINESS WITHOUT PRODUCING IT THAN WE HAVE TO CONSUME WEALTH WITHOUT PRODUCING IT. G.B.S. Above one of the center windows, I read, OF ALL THINGS IN THE WORLD PEOPLE ARE THE MOST PRECIOUS. MAO; and between the next two, WHY DID GOD PUT THE OCEAN SO CLOSE TO THE LAND? Above the far starboard window was THE SPIRIT OF LIBERTY IS THE SPIRIT OF HE WHO IS NOT TOO SURE THAT HE IS RIGHT. JUDGE L. HAND; and, beneath that window, A BOAT IS A HOLE IN THE WATER SURROUNDED BY WOOD INTO WHICH ONE POURS MONEY.

I had just finished reading the inscriptions on the wall beyond—KNOWLEDGE KEEPS NO BETTER THAN FISH. E. M. WHITEHEAD, and, below that, THERE ARE LIARS, DAMN LIARS, AND EXPERT WITNESSES. C. DARROW, and THERE R 3 KINDS OF MEN, THOSE WHO R ALIVE, THOSE WHO R DED, THOSE WHO R AT SEA. B. W. SINCLAIR—when I heard John come in, sliding the green door of the pilothouse shut behind him.

I turned around. On the inside of the door were the words CLOUD WATERS—SKEENA. "What does that mean?" I asked.

"The Skeena is one of our tiptop salmon-spawning rivers, in northern B.C.," John said. "Skeena is the Indian word for 'cloud waters.' The Skeena is misty, with huge dripping trees along its shoreline. I think it's a very beautiful name for a beautiful river and I wanted it here. Did you see 'Lord, Lord, wasn't that a fish. Price of everything, value of nothing,' when you came in? That's by Oscar Wilde, and one of my favorites. I write on the walls with this," he said, holding up a laundry marker he took from one of the three open shelves above the sink.

John pumped water into a dented chrome kettle spattered with white paint, leaned over, and opened the oil valve at the left of the stove, raised the lid on the left corner of the stove top, and lit the burner inside the pot with a match. He set the lid back in place, put the kettle over it, and said, "I've had five boats. This one was built here in the harbor for me in 1952 by a local boat builder named Fred Crosby, and in 1972 my friend Jimmy Reid and I lengthened the *MoreKelp* four feet at the stern to forty-one feet, the way you see it now, and rebuilt the pilothouse, here at Jimmy's. I lost a hell of a lot of my better sayings when we changed the pilothouse. I really miss the one about Samson slaying ten thousand Philistines with the jawbone of an ass. These sayings go back years. I just love whatever I put down. Look at these two, for instance." He pointed above the near window to PUT BRAIN IN GEAR B 4 MOUTH IS IN MOTION, and, 2 SOON OLD—2 LATE SMART. "Something like that applies to me all the time."

He reached for a jar in the cupboard under the sink, spooned tea from it into a brown crockery teapot, and set it on the upper right corner of the stove. "This is a wonderful marine stove—a Dickinson," he said. "The hottest spot is on the lid over the oil burner I just lit, and this far side keeps things warm. These stoves cost almost a thousand dollars now, I guess—they have to be perfectly calibrated to keep going in rough weather—but I bought my original one from Old Man Dickinson himself for less than a hundred. I had to buy a new stove top last year, and it cost more than the original stove. When I used to go into Dickinson's shop to get a part, he would always say, 'Don't tell me how good my stove is. Tell me what's wrong with it!' "

I had resumed my wall reading, and WATCH FOR A GIGANTIC HOAX, above the sink, puzzled me—also another one nearby: SHOOT 'IM WITH YOUR SPORTS JACKET. "There was a guy who sat in the back of the electrical repair shop of a friend of mine in Coal Harbour in Vancouver in the winter of 1936, who prepared coffee and never smiled," John explained. "In the middle of your conversation, he would sigh and say, 'Yes, life is a gigantic hoax.' 'Shoot 'im with your sports jacket' refers to a story that appealed to me about an Eskimo who sold his gun to a tourist in exchange for his sports jacket. In winter, when the Eskimo and his family ran out of food, he said he wanted to go hunting but had no gun. So his wife said, 'Shoot 'im with your sports jacket.' "

The water was boiling, and he poured it into the teapot, stirring the leaves vigorously with a spoon. He filled two mugs with tea, added dark rum out of a bottle from under the sink, handed one of the mugs to me, and sat down on the bow seat with the other, with his back against the steering wheel, facing me.

"I picked up the saying about the spirit of liberty by Judge Learned Hand a long time ago, from an article I liked in an American magazine," he continued. "The one about consuming happiness without producing it is among several I've got on these walls by George Bernard Shaw. I read his prefaces particularly. I love that inscription about lawyers shoveling smoke by Oliver Wendell Holmes. Lawyers are a real excrescence on society, apart from a few labor and civil-liberties lawyers. An overly big proportion of the wrong kind show up in politics with nothing to say. Because they are taught to speak well, the sheep fall for them. Most people who do have something to say don't have to learn oratory. If they feel strongly, they do a damned good job. How do you like this one behind you?"

I turned and read, THROW AWAY TV AND START READING. "I invented that myself," he said. "TV is one of the best media and the one most misused to push violence and hatred."

He looked at me quizzically. "I like to be reminded of things I have read or heard or thought about that meant something to me, so I write them on the walls. I guess it's all a bit different from what you expected."

"I didn't know what to expect," I said.

"The *MoreKelp* is a good troller—exceptionally so in the trough of the sea, with a nice slow roll. That's what counts," he went on. "I spend hours trolling. It's better to have those hours in comfort than the ones when you are running at full speed. This is not a good sea boat; it's a narrow, lousy one in a following sea, which is when the wind and sea behind it push your boat from side to side, off course. At times like that, it would be better to have a wider, shallower boat, but then its rolling action would be too fast for trolling. It's very deep for a troller—about six and a half feet. It's forty-one feet long but only ten feet wide—a foot and a half narrower than the average fishing vessel." He looked around and sighed. "The pilothouse is sixteen feet long and six wide, and designed so I can fish alone." He waved his hand at the stove and steering wheel. "I cook my dinner with my right hand while I'm steering with my left. I often wish I had made the pilothouse a foot wider. It would have been a *lot* more comfortable. The *MoreKelp* has a reputation for catching fish, but it is the single most uncomfortable fishing boat in British Columbia. No doubt about it."

II

I returned a month later. Early in the morning on Boxing Day, which is the day after Christmas in Canada, I followed John, who had a red stewpot under one arm and a box of supplies under the other, down the hill from his house to his dock and aboard the *MoreKelp*. He lit the stove and put the stewpot in the oven. Then he started the motor, cast off the lines tied to the scow, lifted up the three red balloonlike fenders, which hung from the boat on heavy ropes at intervals over the side to protect it from rubbing against the scow, and dropped them back on the deck. "These are called Scotchmen, because Scotland's herring fishermen used round canvas floats filled with air to float the ends of their nets," he said. "Now the floats are made of plastic. We use them for fenders, as marker buoys, and in a pinch even as life preservers."

We headed for the open water outside the harbor. Standing out on the stern deck, I watched John's house—the weathered silver-gray shake roof, the broad unpainted brown side planks, the light yellow trim around the front windows—disappear as the boat carried us away. The house, the workshop down by the rocky beach, the crooked wooden ramp, and the scow were lost in the trees as we turned right, around a point and toward the harbor's entrance: a journey of twenty minutes at our slow speed, which took us past islands, coves, and bays like ours. The sun shone, but a stiff wind beat on my red windbreaker, so I went into the pilothouse. John, smiling from the big seat where he was steering, leaned over and unhooked the backless stool. I sat down on it at the right of the wheel, which turned slowly from side to side as he steered casually with one hand. He was wearing his heavy brown wool pants, thick yellow leather ankle-high boots, and a rather handsome gray knit sweater, open at the top, with three white buttons mounted on a gray cotton binding that went all around the neck.

"What a great sweater!" I said. "I've been looking for a gray one like that."

He stared at me. "You're joking," he said.

"I'm not," I replied.

John burst out laughing. "I was about to apologize for my costume, but these are good practical fishing clothes and what I always wear," he said. "This sweater you admire is a Stanfield's underwear top. I don't suppose you see Stanfield's underwear in the United States, but it's famous in Canada. Most men wear a shirt *over* it, but I like it the other way around. When it gets full of blood and grease and everything else that goes with fishing, I just put it in a bucket of water and wash it out. Besides, it fits down over my hips and keeps me warm. Very good for my arthritis." Still laughing, he added, "If I can't find a Stanfield's small enough to fit you, I'll give you one of mine."

We were traveling slowly through the harbor, past the pink buildings of the Garden Bay General Store, past houses with sun decks facing the water, cantilevered from the sides or on top of the cliffs. No sandy beach here; nothing but sheer rock piles falling abruptly into the sea. You could identify the summer people and the newly retired by the sameness of their expensive houses on narrow lots, their picture windows and sun decks with sliding glass doors; the neatness of their docks, with painted iron ramps, blue, white, brown, or green, connecting to smooth concrete abutments on the shore. The smaller, irregular-shaped dwellings of old-time residents, still holding on to their land, had more waterfront; and often had haywire dock arrangements like John's which might date back to the nineteen-forties or early nineteen-fifties, before there were roads and when everything came in by boat. Interspersed among these were occasional government docks, marked by creosoted pilings, red ramps, and broad floats surrounded by a haphazard grouping of large and small vessels, representing every kind of water activity: fishing boats, fish packers, logging workboats, police cruisers, yachts, sailboats, dinghies, and outboard-motor boats.

Little pleasure craft ran past and around us, one coming so close that John leaned out the window and shouted at its operator. In reply, the boat shot across our bow, barely missing us. John suddenly swerved, steering directly at the smaller craft, and ran it so close to the rocky shore that I gasped, startled by John's quick fury. The pleasure boat, now frantic to get away but caught between us and the shore, was headed for the rocks when John abruptly turned the bow straight ahead again. He leaned across me, released a leather strap at the bottom of the right pilothouse window which lowered it, stuck his head and shoulders out the opening, and shook his fist at the offending craft, shouting something that was swept away in the noise of the motor and the wind. He sat down again. "That bastard was violating every rule of the road—not keeping to the right, and going so fast that when the waves from his wake hit the shore they are likely to cause small boats to crash against their moorings, the docks, and one another. Damned inconsiderate sportfisherman! I hope he's learned his lesson, but probably not!"

We were passing a handful of uninhabited islets with picturesque ruins—unpainted remnants of frame shacks with collapsing roofs and gaping windows. "Indian reserve," John shouted. "The Sechelt Indians used to summer here from their big reserve farther down on the peninsula. These are probably the most photographed islands on the coast, and artists love to paint 'em." On our right, on the sloping lawn of a large white frame house, was a white plaster statue of a bearded man whom I presumed to be Neptune because he wore a long, draped robe and carried a staff. John slowed down to pass a fisherman on our left who was standing up in a dinghy, holding a fishing rod in the water. The man looked up and waved, and John leaned out the port window to shout, "Got anything yet?" The man put up two fingers and smiled as we passed him, his boat bobbing up and down in our wake. "That's Bill McNaughton. He and his brother, Don, were brought up in Jervis Inlet, where we're going now. They lived there for years before moving to the Harbour, trapping, and hunting seals, for which the government used to pay a bounty."

John reached over to push the throttle, one of two brass handles—the other was the clutch—on the shelf in front of me, and picked up speed as we hit the wide-open, white-capped water of Malaspina Strait. For a moment, we were facing the long, mountainous land mass, Texada Island, that blocks the view from John's house of the snowy peaks of Vancouver Island on the other side of the Georgia Strait. Then we turned right into Agamemnon Channel, a passage nine miles long and a mile wide bordered by tree-covered cliffs on both sides. John motioned for me to sit on the seat beside him. He got up to let me by, then pulled down a large, desklike board hinged to the ceiling which was full of charts, selected one from the pile, and moved it to the top, tracing with his finger the fjord winding to our destination, Deserted Bay, fifty miles away, almost at the closed end of Jervis Inlet. "While I'm there, I'm hoping to go up on the mountain and find two narrow cedar trees thirty or forty feet high that I can plane down for fishing poles for the boat," he said. "The ones I've got are developing deep cracks."

On the chart, he showed me little numbers in the water areas which indicated depths in fathoms—six feet to a fathom—and said, "I constantly look at my charts when I am traveling, and even more when I'm fishing. It's the best way I know to keep off the rocks." Land was colored a light tan, water white, except for small patches of blue denoting shallow water, six fathoms or less, along some shores. Where we were going, the numbers varied from fifteen to twenty fathoms into the hundreds of fathoms, even close to shore; and down the middle of the channel in some places it was over three hundred fathoms. "It's really deep in here," John said. "Always remember that the *MoreKelp* probably won't run aground between two and three fathoms, but you're safer at six."

Moving along to the steady thrum of our motor, we seemed to be going directly into the snowcapped mountains, as if we would be able to step out on their peaks when we arrived. At noon, John turned off the engine, and we drifted while he made a quick lunch of tea and large disks of a Scandinavian-type crisp bread called Britl-Tak, which he broke into pieces and spread with thin layers of margarine, peanut butter, cheddar cheese, and, finally, slices of cucumber. He pulled down a smooth board hooked to the wall, whose other side had a raised rim and was covered with a green plastic mesh mat. This was our dining table, midway down the bunk, and we sat down on either side. In the narrow space on the wall behind the table, four shelves had been built which held two glass egg cups, a small collection of eating utensils in a plastic cup; cans and jars of seasoning and spices; a Pyrex pie dish that he placed on the mat for me; and a chipped white enamel plate pocked with black areas of tin showing through, for himself. "My favorite dish," he said. "I had two of them, and one went overboard in a high wind. I almost sank the boat trying to get it back. If I lose this one, I might have to stop fishing!"

For dessert he spooned out a blueberry-and-apple mixture on our plates from a container he had brought among the supplies from home.

"The apples are from my old tree below the house, and I picked the blueberries in the early fall, cooked up a mix, and froze it," he said. "I can hardly wait to take you blueberry-picking up the mountain; the views are *fan*tastic, and it was so warm and sunny and lovely the day I picked these. I stayed up there until I could barely see. I love blueberry-picking and try never to let anything interfere with it. Getting up on the mountain is very important in my life and is part of what I am all about. On fall days, as soon as the sun shines, I have a bugger of a battle, a pull thisaway and thataway—between my desire to hike and ski madly and the bloody boat chores—especially painting, which can't be done in the rain. I love picking in October, with its lovely fall scents and crisp leaf-changing-color days. At twenty-five hundred feet, the air is like wine."

We kept plucking out leaves and twigs until we had a small pile of them on our plates. John laughed. "I use the flotation method to take the leaves and twigs out. To tell you the truth, I *like* the crunchy texture."

After lunch, John took out a battered round aluminum dishpan from a narrow shelf under the stove just above the floor, put the pan in the sink, and poured hot water from the kettle into it. He handed me a bottle of liquid soap, a tea towel, and a dishcloth that he pulled from an open round hole in the ceiling. "That hole is for ventilation, but I stuff it with the dishcloth to hold the heat in the cabin when the stove is on," he said.

John turned on the motor, and we were traveling again. I washed and dried the dishes, put our eating utensils back in their slots, and hitched up the table, closing in the dish shelf and the condiment shelf behind the table. Then I hung our cups on their hooks above the sink.

"Don't do that!" John shouted.

I jumped. "Do what?" I said.

He was now standing by my side, still holding on to the steering wheel with his left hand, looking really upset, and I didn't know why. "My God, don't you see what you've done?" he cried, taking down the two cups I had just hung up. "You had them facing in opposite directions and opposite to the two cups already there. Start right now rehanging them all in the same direction." He paused as I slowly put them back facing one way. "What do you think would happen at the first roll of the boat in heavy weather? They would hit against each other and smash to bits. Think ahead before you do *anything* on a boat."

I must have looked as frightened as I was, because he put his arm around my shoulder and added, "Wait until we've been ten days on the boat in roll, roll, roll, rain, fog, and cold, and the frying pan has bounced off your feet for the fourth time and the stove won't heat and the vegetable pot heaves into the sink just when you are ready to eat! That's all right. You'll forget the roll when you see one or two of our northern rainbows and sunsets."

We went back to our perches on either side of the wheel as the boat chugged deep into the Inlet. Always, as the channel turned, the mountains seemed to turn, too, and encircle us, reappearing at new angles, white and snowy, just ahead in a different setting. Sometimes John would steer the boat toward the steep cliffs to look more closely at the churning white foam of a waterfall whose top edge would be lost in mist; or we would come out in an enormous bay that took thirty or forty minutes to traverse.

At about two o'clock, John started yawning. He said, "Here. You steer. Just keep to the center of the channel slightly toward the right. I'm going to take a nap." Before I could protest, he had flopped down on the bunk with a gray wool sock over his eyes and appeared to fall fast asleep.

I gripped the wheel, which was covered with a thick coat of aluminum paint. It felt lumpy in my fingers, and cold; or was it my hands that were icy? Tentatively, I moved the wheel from side to side as I had seen John do. The swift response of the boat shocked me, and the shivering needle on the large compass before me on the shelf mirrored my uncertainty. I tried to think: What could I have said to John that would have given him the mistaken impression that I knew how to steer a boat? I turned around to see if he was really asleep; he was snoring softly. Well, I thought, this is certainly a case of misplaced confidence, but I might as well try to enjoy it, so I settled down into a sort of steering rhythm that produced an alternating movement of the compass needle steadily between NE and NW. This seemed to hold the boat to the center of the channel, more or less, and I was able to look at the scenery: the steep cliffs topped by green coniferous trees, with an occasional stand of the shorter arbutus, or madrona trees, that I could recognize by their reddish-brown bark and the graceful twist of their smooth trunks, growing right at the edge or down the embankments.

The land looked deceptively close but was probably a mile or two away. I made a game of seeking out half-hidden cabins among the trees or searching with John's binoculars for the heavy trucks and tracked vehicles parked around groups of buildings at the water's edge on raw strips of road that wound up to stands of timber, indicating a logging operation. I thought: What a huge effort it must take to keep these small clumps of humanity going when everything, people and equipment, has to be brought in by boat! I searched for human figures moving about; aside from a single smoking chimney, the whole coast seemed abandoned, except for us.

A momentary gush of terror: a boat coming toward us. It passed on our port side by a wide margin—probably a mile—but my heart was beating so rapidly I could hardly breathe. It reminded me of how I felt the first time I passed another automobile when I was twelve and learning to drive, propped up at the steering wheel of my parents' Franklin car, with three pillows behind me. Looking at my watch, I realized we had been running for almost six hours and the sun was going down behind the mountains that were darkly closing in around us. John woke up, took over the steering, and we were in a large bay, passing an empty black barge anchored out from shore, moving slowly toward a dilapidated half-submerged ramp. "This is Deserted Bay," John said. "Harry Dray and his girlfriend, Minnie, have seen us approaching and will be coming down the hill any minute. This used to be a logging camp. By law, logging concerns have to burn everything when they pull out, but Harry had been the caretaker here, and he and Minnie wanted to stay on. They bought their cabin from the owners of the camp, and Harry uses the barge as a breakwater."

Sure enough, a man astride a small red tractor was descending the hill, followed by a short, stocky woman, walking. "Harry's had several bad heart attacks, and each time the local airline has had to fly in to bring him out to hospital, but he still goes two miles across the bay every day in all kinds of weather, winter and summer, in an open boat to caretake a logging camp on the other side," John said. "Minnie's Norwegian, and she and her sister, Bergliot, were brought up to log and hunt like men by their father. Bergliot lives down the coast not far from us, and they both have registered traplines. Minnie likes to go out at night and hunt, and sleep in the boughs of trees. It takes really strong people like Minnie and Harry to stand this place all year. The mountains in these inlets can press you down, along with the snow, ice, and waterfalls. The pot of stew I brought is for our supper with them, and the bag of poultry feed on our deck is for Minnie's ducks and chickens."

We eased into the dock, which was listing heavily to one side; cut off our motor; and John threw a rope from the bow to Harry Dray, who tied it to a rotting post. Harry's face had a chalk-white pallor, but Minnie, standing beside him nodding and smiling, was glowing with health. When I climbed over the side of the boat and shook her hand, it was as wide and rough as a man's. She was about my height—five feet two—but twice as broad. Everything about her gave an impression of roundness: her face, her body, her short, powerful legs, even her feet, in men's shoes. Her long, brown hair under a man's torn felt hat fluttered with each bobbing of her head; but her most distinctive feature was the leather belt that held up the men's dark wool pants like John's which she was wearing. It was two inches wide, studded all the way around with shiny bullets, had a huge metal buckle, and bristled with hunting knives. John handed her the bag of feed, which she easily swung over her shoulder, striding up the hill; John and I walked behind her, and Harry came along beside us on his baby tractor.

On the way, we were greeted by several dogs, short-legged and round, and the cacophony of a heterogenous collection of live poultry, cackling, crowing, and honking from small pens separated by fishnet, boards, and rusty bedsprings set on their sides, connected by chicken wire. There were chickens, ducks, and geese, and, in one quiet pen, white and tan rabbits. Minnie beamed like a proud mother, showing off her Peking and Muscovy ducks. At Harry's insistence, we detoured into a shed, where he turned on a noisy generator, and we looked into a large trunk freezer stuffed with berries and vegetables from last summer's garden. A freezer in Deserted Bay! Later, I found out that even though the generator didn't run all the time, the electricity was on long enough every day—usually in the evening, when electric lights were desirable—to hold the freezer's temperature to below thirty-two degrees.

We left our boots in the vestibule of the log house. Externally it had the sagging characteristics of a shack, but, upon entering, I was startled to see that it was quite sturdy and had electric lights, a refrigerator, and a flush toilet. Exotic amenities in this isolated place! Minnie closed the door behind us, exposing a shotgun and three rifles on its interior side. John placed our red stewpot—which contained a pork roast that had been cooking as we traveled—on Minnie's stove. He pulled a gray wool sock with a bottle of Scotch in it from his jacket pocket and set it down on the table. Minnie opened a door in the side of the stove, slipped in a stick of wood, and put potatoes and carrots to boil in a pot on top. From a cupboard, she brought out a loaf of homemade bread and a jar of honey from their own bees, while John carried in two chairs from the other room. Minnie motioned me to the one broken-down, overstuffed seat by the stove. Harry had already settled into a straight kitchen chair in front of the oven, opened the door, and propped his right leg on the rack inside. "Have to keep my leg warm, John. I don't have any circulation left in that one," he said. "They had to send the plane in and take me to hospital once already this year, but I'm doin' O.K. I had quite a time here this fall." He grinned. His white face in the dim light of the electric bulb had taken on an eerie bluish tinge. "You know that camp I caretake across the bay? Well, during a terrible storm last October, the engine broke down on my kicker on the way over and I had to row two miles in a horrible sea without oarlocks."

"Jeez!" John exclaimed, wagging his head in amazement.

"Yup," Harry said, still grinning. "Collapsed on the dock when I got there."

"You got any oarlocks now, Harry?" asked John.

Harry slapped his bad leg and laughed. "Nope. Not yet. But I may get 'em soon." He paused. "How you feelin', John?"

"Fine, but I take care of myself," John said gruffly. "How's the hunting, Minnie?"

"Yah, pretty good," Minnie said. She spoke with a thick accent, ending each sentence with a nervous giggle and bobbing her head up and down. "I shot a real big black bear last shpring up there in our woods. It were a real heavy bear when I went to get it, and had such turrible worms I just shkinned it and brought back the hide. I took it down to my schister and put it in her freezer, but schomebody tripped over the electric cord and the freezer went out. That was schome untreated bear hide! It sure schmelled! Finished. That's all right. I get another."

"Hunt, hunt, hunt. That's all that woman ever thinks about," Harry said.

The long trip up the Inlet had tired me, and I fell asleep in the big old chair after supper, about seven o'clock, with one of Minnie's cats on my lap. I didn't wake until it was time to stumble sleepily down the hill to bed. I slept in a sleeping bag on the narrow board floor of the pilothouse, between the bunk and our oil stove. It was so comfortable, so warm, that I didn't open my eyes until six the next morning.

John's bunk was covered neatly with the canvas, and he had disappeared. I looked out the pilothouse windows to see where he was.

A film of light snow covered the deck, the ramp, an old fishboat and dinghy over by the shore, the gentle rise of the hill, and the roof of Harry and Minnie's house, from which I could see smoke curling out the chimney. The mountains around us, bathed in a lavender light as the sun rose, were covered with snow that gave the trees close by a woolly look. I rolled up my sleeping bag, washed my face in cold water from the pump, brushed my teeth, and was making tea when the door opened, bringing a rush of crisp, cold air.

John entered, stamping the snow off his feet, which were encased in hip boots. He was carrying a galvanized-iron pail full of fresh oysters he had picked off the rocks at low tide while I was still asleep. He changed from his waders to his yellow leather boots while I was preparing breakfast. As soon as we had finished eating, John went up the hill to meet Harry, whom I could hear starting the motor on an old pickup truck, to drive him up the mountain to look for trolling-pole-size cedar trees. John was in a hurry; he wanted to start back in plenty of time to get home before dark.

The night before, at my suggestion, John had negotiated with Minnie to buy one of her ducks for roasting, so after I had cleaned up the dishes and swept the toast crumbs off the stove and floor of the pilothouse with a whisk broom into a dustpan, I walked up the hill to pick up the duck. Minnie was standing by one of the pens waiting for me, holding by its feet a headless duck dripping red in the snow.

We made another tour of the pens, then went into the house, where she wrapped a piece of newspaper around the feathered body of the duck. One of the Muscovy ducks followed us right to the door, its red wattles quivering, making a sound between a wheeze and a honk which struck me as particularly heartrending. When we came out again, he was wandering up and down, almost pacing, at the open door of his pen.

"Is he a special pet?" I inquired.

"Ach, no, he's just feeling bad. He's lost his friend," Minnie said, handing me the package with the duck in it.

John returned three hours later from an unsuccessful search for cedar saplings straight and tall enough for trolling poles, and we set off immediately for home. "I would have preferred to get new poles from at least fifteen hundred feet high up on the mountain. Under that, they are usually full of powder worms that bore into the wood and make tunnels, but I'll find something in my own woods that will do," he said. "I got the mast for the *MoreKelp* from a tree right outside my sons' bedroom window at home—an old snag of a tree with a white top that bald eagles and blue heron liked to sit on. We call that kind of a tree a 'widow-maker.' It's a dead tree that might look O.K. but can fall down any time. It turned out that this one was different. It had been killed by fire, which meant it didn't rot, and, by Jeezus, when I swung at it with my double-bitted ax the ax sprang back like a yo-yo. I had to use my power saw instead, and the tree actually bounced when it fell. It was hard and springy and light, ideal for a mast, so I wrapped it in gunnysacks to keep the sun from splitting it, and worked on it off and on for four years in my boat shed. It was the central length I wanted, and when I started planing it down it was so big I could hardly get my arms around it. When I finished planing, I couldn't quite circle it with my hands. It's second-growth cedar from about 1900—really strong. It supports the heavy poles, guy wires, and all the rigging that hangs from it, and it has to be able to give when I'm towing huge weights, like poles and leads and stabilizers. It's got dimes and fifty-cent pieces underneath. Nobody knows why, but it's a tradition among fishermen to toss silver down under the butt when they are raising the mast."

I steered while John slept. When he woke up and took over the steering, I tried to take the feathers out of the duck, but I didn't know how, or it was an especially tough duck; either way, my heart wasn't in it. I put it aside and went on deck and shucked oysters, eating one for every two I dropped in the pot. "Eating raw oysters!" John exclaimed, coming out of the pilothouse. He walked past me to the stern, climbed across the empty fish box, dropped into the cockpit—the standing space behind it—and leaned down to pull the throttle back and reduce the boat's speed to what would be a very slow walk. Then he moved over to a large spool with a crank mounted on a temporary plank that he had placed across the fish box. "This is a hand gurdy," he said. "The main fishing line that is wound around it is made of braided steel wire." He showed me that the end of this wire on the gurdy was weighted down with a lead ball resting right now in a bracket on deck. Next, he prepared a length of monofilament—clear plastic line—which he called a "leader," with a "lure" to attract the fish. Using a pair of pliers, he crimped a black hook to an egg-shaped piece

We were anchored up the coast for the night after fishing, and John was relaxed and reading, when I made this pencil drawing of him.

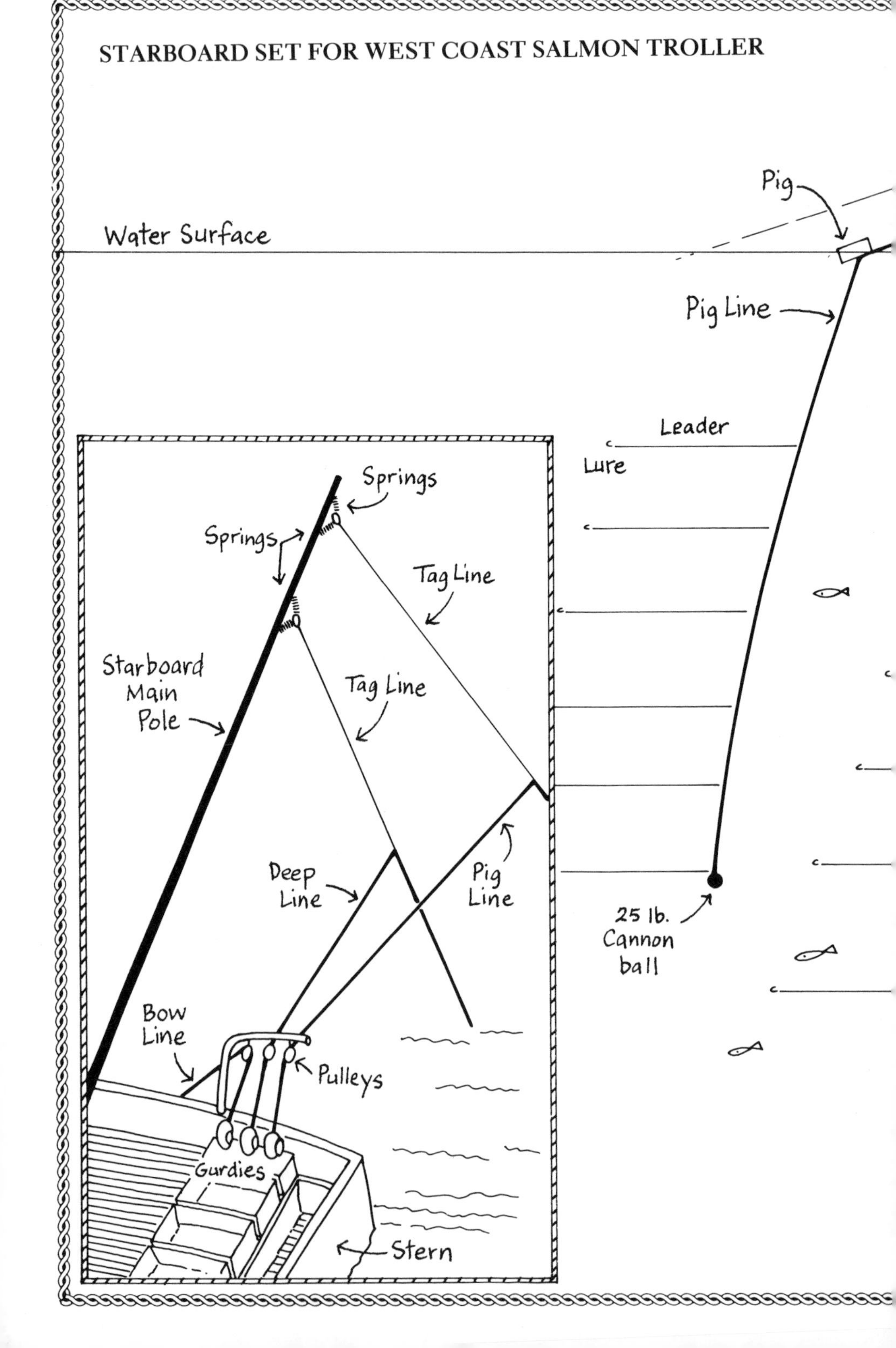
STARBOARD SET FOR WEST COAST SALMON TROLLER
Water Surface
Pig
Pig Line
Leader
Lure
25 lb.
Cannon
ball
Springs
Springs
Tag Line
Tag Line
Starboard
Main
Pole
Deep
Line
Pig
Line
Bow
Line
Pulleys
Gurdies
Stern

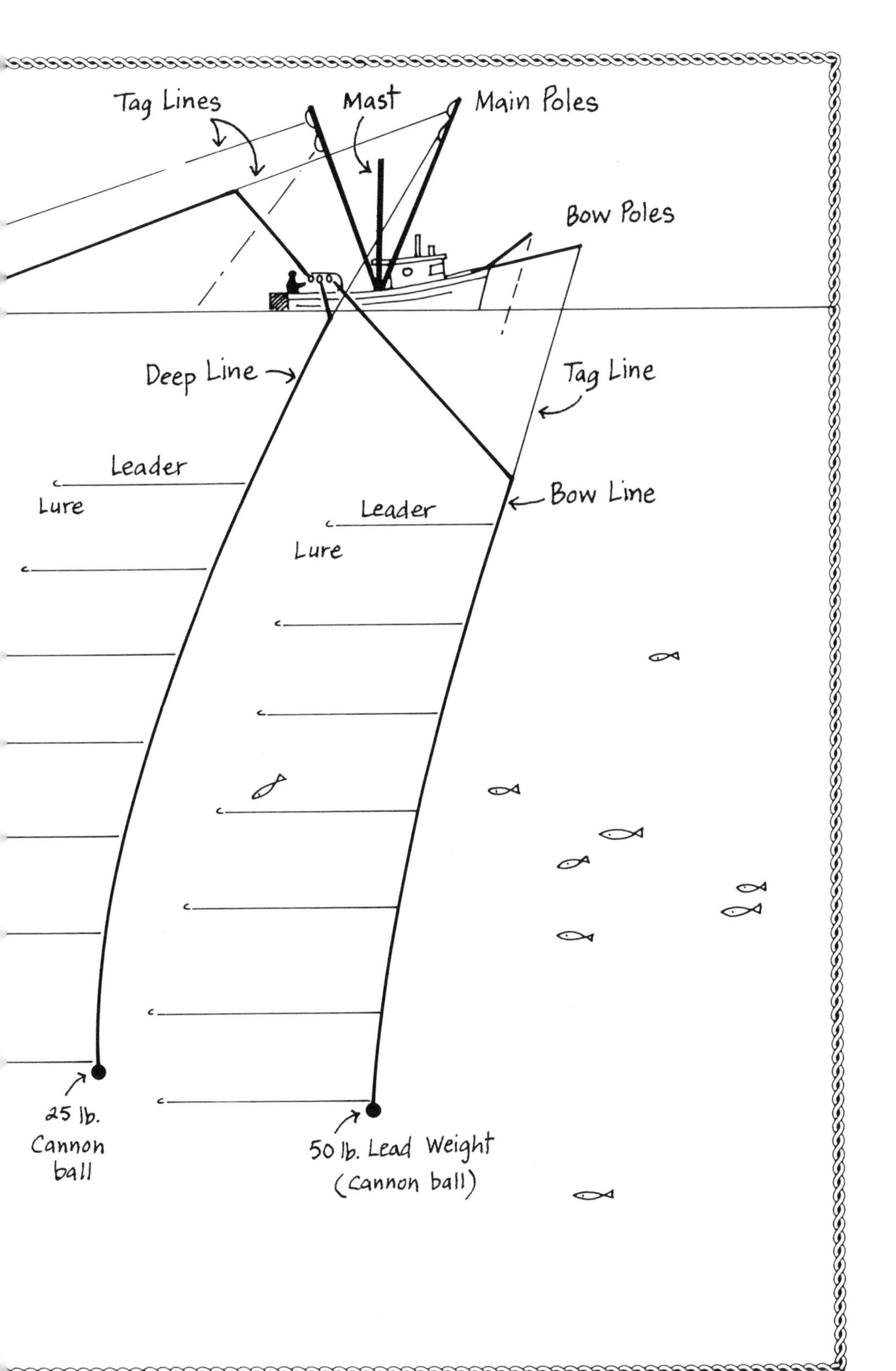
Tag Lines
Mast
Main Poles
Bow Poles
Deep Line
Tag Line
Leader
Lure
Leader
Bow Line
Lure
25 lb.
Cannon
ball
50 lb. Lead Weight
(Cannon ball)

John displaying the forty-pound spring salmon he had just hauled in with his gaff. Springs were his specialty.

of shiny brass plate that he called a "spoon," which he then tied to the leader, explaining that when he clipped the leader into the steel mainline it would do just that—lead the line, with its lure and hook, down to where, one hoped, the fish were.

John lifted the lead ball and had me hold it. It was really heavy. "This cannonball weighs twenty-five pounds," he said, clipping the leader in and dropping it over the side. The steel mainline spun out from the wheel, or gurdy, in pursuit of the lead, with lure and hook attached. It immediately sank below the surface. "It needs to be heavy to hold the lines and hooks straight down," he added. He let the wire spin off the spool for what seemed like several minutes before he braked it. "We're down thirty fathoms now," he said. He sat then with his hand lightly touching the line near the top of the spool wheel and waited.

I offered him a raw oyster, but he rolled his eyes, shivered, and refused it. "I don't eat raw oysters," he said. "It's not safe. Oysters may be all right, but an awful lot of fish have worms." I had never tasted oysters so fresh, and I wasn't listening to John; I only heard what he had said later, as in a dream.

Several times, John wound in the line close enough to lean over and examine the hook in the water, grimacing and grunting from the effort of cranking it up by hand with that heavy weight attached. "In winter, we have to take off all the commercial gear, including our power gurdies, which run from a hydraulic pump off the engine. We can only fish for sport, with one handline, and aren't allowed to sell what we catch. The fishing season for us commercial trollers doesn't open until April. It sure is hard work, pulling up a twenty-five-pound lead this way, but it's worth it if we get a salmon." He leaned back, his face lifted to catch the warm rays of afternoon sun. Before leaving the pilothouse, he had turned on the radio to his favorite program, the Canadian Broadcasting Corporation's Music Hour, which was on from two-thirty to three-thirty weekday afternoons, and switched the sound outside through an aluminum loudspeaker encrusted with salt and corrosion, attached to the back of the pilothouse above the anchor winch.

The strains of a Mozart sonata, a lovely sound even through the tinny old speaker, mingled with the soft hum of the engine moving us ever so slowly through almost-still waters. We were in a broad stretch of the Inlet, the cliffs on both sides at a fair distance, with no other boat, no other human beings to disturb our tranquillity. The sun glowed golden on the mountainsides, and we took off our outer jackets, basking in the warm afternoon rays.

John stood up suddenly and began winding in the fishline, but almost immediately there was a look of acute disappointment on his face. I saw a plump reddish-orange fish swing over his head on the line, followed by a thud, a whack, and then stillness. I looked at the fish in the fish box. It had huge, bulging eyes, a brilliant orange-red head, and spiny fins that hung like drapes from its rounded body. It would have been beautiful if its tongue and a floating bag of whitish material had not protruded from its capacious jaws. While John was disengaging his hook from its mouth with pliers, I said, "What kind of a fish is that?" and, before he could answer, "What's coming out of there?"

"It's a red snapper," John replied. "We sometimes call this fish Norwegian turkey, because it was a preferred delicacy among early Scandinavian settlers. Norwegians *love* cod, and this is a form of it, and a bottom fish. The pressure of coming to the surface from so great a depth ruptures its swim bladder and pushes its stomach out. It won't survive if I throw it back, so I'll take it home and put it in the garden for fertilizer. We could eat it, but I hate to fillet snapper—all those lethal spine bones—and I do it very badly."

He had already tossed the line back in the water, and after a brief interval he was furiously cranking it in again. "I think we have a real bite this time," he said, and then he was winding in the line, looking delighted. I saw a flash of silver and a thrashing tail, as he landed the salmon in the box. "A small spring," he said, picking it up gently and holding it against a crisscross of measured lines carved into the trough that bordered the stern side of the fish box. "Nineteen inches—big enough to be legal for sportfishing, which we are doing now. It's twenty-six inches when we fish commercially. We'll take it home for supper." He took up a white stick shaped like a baseball bat, with a heavy hook on the narrow base, lifted the fish on the end of the line to which it was still connected, and whacked it hard on the back of its head with the broad top of the stick. "I kill them right away, because I don't like to see them suffer," he said. He removed the hook from its mouth and dropped the fish into the fish box, where its tail flipped for several minutes—violently, less violently, then stillness.

I picked up the death instrument and turned it over in my hand.

"That's called a 'gaff,' " he said. "The hook end is indispensable for hauling in big fish. I made it myself from a baseball bat."

He took a long, sharp knife to which a metal spoon was attached at the reverse end of the wooden handle, and, holding the fish firmly in the trough with his left hand, made two swift cuts with his right—the first, a circular scoop that removed the gills; the second, a slit up the belly from the anal vent to one and a half inches below the gills. He cut its diaphragm with the tip of the knife, reached inside with his rubber-gloved hand, and ripped out the internal organs, throwing the eviscerated parts overboard. Raucous calls from the sky above, and a flock of seagulls swooped down to the water, seizing the remains in their beaks. John turned his knife to the spoon end and scraped out the blood from the backbone. All this took less than a minute. He leaned over the side and dipped up a pail of water and washed the fish, inside and out.

He dropped the fish, with its head still on, back into the fish box, covered it with burlap, and put the gear away. "One is enough," he said, climbing out of the stern and returning to the pilothouse as the sun went down. "All I wanted was one salmon for our supper. It's a nice, fat white one. We fishermen think white spring salmon are the best; they've got more fat and have a richer taste." The term "spring" mystified me; I had never heard of that variety of salmon. "A spring is what you Americans call a chinook, or king, salmon," John explained. "I don't know why we call it a spring. Probably because it is the first salmon to show up in the spring in our waters. But there are spring springs, and springs right through into fall and winter. Which makes it a little confusing, I suppose."

As soon as we got home, John finished cleaning the duck and I put it in the freezer in the basement. Then he rolled the salmon fillets in cornmeal, and fried them gently in oil. The meat was cream-colored, almost white, with the faintest pink tinge. "Lovely, lovely," he murmured as he put the cooked salmon on our plates. He lifted the first forkful in the air. "Nothing in the world tastes better than a freshly caught white spring salmon!" he said, and popped it into his mouth.

III

I went back to the United States shortly after the trip up Jervis Inlet. When the fishing season opened in April, John began writing me a continuous letter on a pad he kept on the compass shelf of the *MoreKelp*. He would tear off letter sections and mail them to me in New York whenever he brought his fish to Vancouver or a more northern port to sell. In one of these segments, he wrote:

> Re: trolling: few who haven't been to it and at it realize what a knock'em-down drag'emout business it can be—from 9 to 11 daze U start that thumping engine at 3:30 to 6 AM & shut it off at 5 to 11 PM—run out 1 to 2 hours & then down with the gear & run slowly (engine then is *very* quiet BUT IT'S ON & that can get U) & U drag gear around at approximately 2 miles an hour for 12 hours & then run into harbour 1 to 2 hours & U just do this for 9 to 11 days & then a 6 to 16 hour run to Port Hardy or Namu, a mad rush of marine traffic jam, of fish, booze, gear, etc., get fuel & water, have an oh-so-marvelous shower (or 2!), buy a restaurant meal or 2 (any greasy spoon seems idyllic then for 1 or 3 meals IF U DON'T HAVE TO COOK IT & wash up in *tiny* boat space), it's a lovely change. On the 2nd morning U R away again for another 10 days & U do this endlessly April 15 to September 15 or 20th. By end of August it's a TOUGH IT OUT A LITTLE LONGER CONTEST! . . . but I love most of it!

Another time, after a very rough trip north:

Incident 2 test temper at 7AM! After 2 attempts 2 clean water out of stove I finally got 2 eggs boiled, put one on seat in egg cup (the softest of course), then a log got in a fishline, had to dash out, couldn't C when came in coz rain on glasses—I sat on the bloody egg & it rolled off and bust in my hush-puppy shoes & then dript down into engine. Do U still want 2 come?

And just before my arrival:

Trip ending. Bread all turning GREEN. I'm chewing whole clove of raw garlic. I *crave* it at times. Very good 4 the blood—it purifies. Remember that big, hard-covered bags are a problem to stow away . . . two flexible ones to cover the same "volume" would be easier. Main items 2 bring: bluejeans, warm underwear & sweaters, and your excellent red wind jacket.

IV

I rejoined John in July, at the height of the fishing season, at a place called Namu. The serious business of fishing terrified me. I wasn't worried about living on a boat or being seasick; I was afraid of being a nuisance by getting in the way or falling overboard. I didn't think about the actual process of fishing; I didn't know enough about it to do that. Permission to join John on his boat during the commercial fishing season came from Ottawa in June, but I had to pick up the formal document in the regional office of the Federal Department of Fisheries in Vancouver before I could proceed north. Without that, John would risk losing his commercial salmon-fishing license, in the unlikely but still possible event that a Federal Fisheries officer boarded the *MoreKelp* while we were fishing and asked to see my papers. I arrived in Vancouver with barely enough time to do some shopping for John and go to the Fisheries office to pick up the paper. TO WHOM IT MAY CONCERN, it read, and went on to grant me permission "to act as an observer, along the British Columbia coast, aboard the troller *MoreKelp* owned by Mr. J. Daly of Garden Bay, B.C. . . . and possibly other Canadian fishing vessels . . . for the purpose of writing . . . and research for a book which she intends publishing."

Running around Vancouver on my errands, I nervously calculated the three-hour time difference between the East and West Coasts and reset my watch one hour short. I missed by that hour the departure of my plane and had to wait overnight in a Vancouver airport hotel for another aircraft going north. In addition to my two small pieces of luggage—a duffel bag and a soft black leather case—I was carrying a large shopping bag of fresh fruit and greens John had requested: raspberries, strawberries, peaches, grapes, lettuce, watercress, eight or nine avocados in varying stages of ripeness, and a half pint of heavy cream. In my luggage I also had two globe artichokes and a jar of preserved kumquats as a surprise. I handed the bag of fresh groceries to the desk clerk to put in the refrigerator overnight, and the next morning, when I took back the bag, it was a solid block of ice. He had put it in the freezer by mistake. During the three-hundred-mile trip north, with a two-hour wait at a halfway point for a small coastal float plane that could land on the water at Namu, the bag sat at my feet, defrosting moistly on the various floors.

On the map John had sent me during the winter, Namu was a name on the mainland surrounded by empty land, almost halfway up the B.C. coast—an indentation in a large body of water called Fitz Hugh Sound. He had explained that during the summer it was a fishing center owned and operated by the dominant fish-processing company on the British Columbia coast, B.C. Packers, where fishermen could sell their fish, make necessary repairs to their boats, mend nets, and take on fresh ice and supplies for the next trip out. In the winter, he wrote, it was deserted, except for a crew of eight men with their families, to maintain the buildings.

From the air, skimming above the trees, I could see waterfalls and streams and the raw scars made by logging—huge strips of exposed rock as if the bearded mountainside had been shaved. Namu came into sight as a clump of white buildings covering several acres of waterfront, with NAMU painted in large letters on the biggest roof. There were substantial dock structures, and a flotilla of fishing vessels appeared to be tied up or moving about. Little figures scurried around as we descended and our plane's floats skimmed the water. All I could see was John, arms folded across his chest, a head taller than anyone else, in faded bluejeans, looking thinner than I remembered him. His beard and mustache were as white as the T-shirt he was wearing, bleached by the same sun that had burned his face and arms so brown. Our eyes met for a second as the pilot climbed down on the plane's float and opened the cabin door, and then I turned around facing the ladder, descending backwards, and stepped off a pontoon onto the dock. A clean-shaven man with glasses, about John's age, was standing beside him, eyeing me curiously, and John hastily introduced us. "Meet my old fishing pal, John Chambers," he said.

John brusquely refused Chambers's invitation to join him for lunch at the commissary, and hurried me and my luggage to a white wooden dinghy (painted green inside), which was tied to the dock. Sitting upright in the rowboat facing him, I noticed he was facing forward, rowing with short, rapid strokes. "I don't think I have ever seen anyone row any way but backwards," I said, "although rowing backwards really doesn't make sense."

"Most of us feel it's a good thing to be able to see what's ahead," John replied.

The *MoreKelp* was riding at anchor, its poles pointing skyward, in their running position. From where I sat, watching the boat rock in the light wind, its deck looked unattainable—as high above me as an ocean liner, although it was about three and a half feet. John threw my luggage over the gunwale; and, with John pushing from behind and my feet desperately groping for a foothold, I managed to scramble over the rail and roll like a sack of salt onto the deck.

I looked around hastily. John was busy pulling up the dinghy, and there was nobody else in sight. I thought, I hope nobody was watching us through binoculars. By the time I had collected my purse and shopping bag, John was attaching the dinghy to a pulley that held it upright on the port side of the open deck, and was moving my luggage to my quarters in the fo'c'sle, below the main space of the pilothouse in the bow. To get there, I slid down backwards from the floor of the pilothouse, since there was no step, to a narrow white bench on the starboard side where John had put my bags.

The port side, my sleeping quarters, was fully occupied by two bunks; the lower one was filled with fishing gear, but the upper had been freshly made with flannel sheets, gray wool blankets, and a pillow with a bright blue pillowcase. John called down that lunch was ready. I had put on jeans, red windbreaker, and sneakers for the plane ride, so I was dressed for the boat. Without unpacking, I climbed right back up to the pilothouse proper, where he was serving the special lunch of abalone he had prepared for my arrival—a coast delicacy I had never seen. He had gone out before dawn at low tide and pried a dozen or so of these tough shellfish off the rocks, pounding them into tenderness and half cooking them before I came. He finished frying them, in butter, and we ate them on the deck in the sunshine, listening to Bach on the CBC concert hour, washing them down with cool beer that John had stored in the cockpit. We had soggy raspberries for dessert, and John threw nine black avocados overboard, one by one. As the last avocado disappeared over the side, I clapped my hand to my head and said, "I forgot! I've brought you something that'll make up for the disaster of those frozen avocados." I dashed down below, opened my bag, and pulled out the two globe artichokes in their brown paper bag, and the jar of preserved kumquats from the sock in which I had wrapped it. I brought them up on deck and handed the artichokes to John. "A treat especially for you," I said.

He opened the bag and looked inside. "What *are* these?" he asked.

"Artichokes," I said.

"Those things you eat leaf by leaf that have nothing on them?" he said. "Certainly not!" He pitched them into the sea, bag and all. He wheeled around, peered down at me with a worried look, and gave me a quick hug. "Never eat the damned things. Too much trouble for what you get." I was still holding the jar of kumquats, so I backed up into the pilothouse and tucked them into the low cupboard, never to be mentioned again.

While I was washing the lunch dishes, John appeared with a box of my favorite Bourbon Creme biscuits—an English chocolate cookie with chocolate filling. "Don't eat the whole box at once," he said. John started the motor and hauled the anchor up. We began running out of Fitz Hugh Sound toward the fishing grounds. The three o'clock marine weather report had issued a small-crafts warning and we were coming out into the vast open water of Queen Charlotte Sound when he began looking anxiously at the sky. "I've been traveling along this coast so long, I could probably find my way blindfolded, but I like to check each anchorage," he said, pulling down a chart and examining it. Before we could find shelter, the wind rose and the rain came down—not gently but in huge, pelting drops that the gale washed wildly across our windows, so that we couldn't see even our bow. The gale wind shrieked through the rigging, and although we were snug inside the warm pilothouse, John was alternately looking at the compass and peering ahead for some physical sign, between gusts of rain, of the safe anchorage he was aiming for. We rocked and pitched in the enormous groundswell. This is what a following sea is all about, I thought, as I gripped the edge of the compass shelf to keep from falling off the stool.

The wind died down as swiftly as it had come up. Although the rain continued, we could at last see landmarks familiar to John. He slowed down when we were in the lee of a small island, and, with the bow pointing toward shore, put the clutch into neutral. He was standing inside at the wheel, and he reached over and opened one of the starboard windows. Then he turned and dashed for the door, reaching down before he went through it to pull a chain that engaged the hydraulic motor that powered the anchor winch, and pushing the clutch of the anchor winch just outside the door with his foot as he passed.

Through the starboard windows, I watched him edge his way slowly forward along the narrow foot-wide space between the pilothouse and the gunwale toward the bow. Arriving on the bow deck, he crouched next to the bow hatch cover directly in front of me, with one hand on a steering wheel mounted on the outside wall of the pilothouse. Reaching in through the open window with the other, he pushed the clutch to reverse. Then he gently moved the throttle so that we backed up a little, continuing to maneuver with throttle and clutch back a little, then forward, until he was satisfied with our position. He stood up and shoved the heavy galvanized anchor that I could see hanging, poised at the bow over the roller, into the water, giving the heavy chain an encouraging tug. With a deep-throated groan, the anchor chain clattered over the bow for a dozen or so fathoms, pulled by the weight of the anchor. John gave it a couple of kicks, and it continued to travel down until the sound changed as the shackle joining chain to rope went over the bow roller with a rumpelty-clunk, down the narrow footpath on the port side, unwinding steadily from the winch. The end of the chain went over the side with what seemed to me like a sigh of relief, followed by the smooth-sliding rope, which ran along quietly until it halted, abruptly. The anchor had touched bottom.

The motor had been idling, and again John pushed the clutch, from neutral to astern, the boat moving almost imperceptibly, tugging at the anchor, until he was satisfied that the anchor was holding.

"How far down is it?" I asked.

"Twelve fathoms," John replied.

He came into the pilothouse then and shifted the clutch to move the boat forward and back once more, to make absolutely sure that the anchor had caught and would not drag during the night, and said, "Stop the Gardner."

"What? Where? Where?" I cried. I didn't know what he was talking about.

"Stop the engine. Pull that green rope behind you to the right of the pilothouse door," he said. "It's beside the exhaust pipe."

I ran to the back. An implausibly slim vertical green rope came up through a small hole in the floor there and was attached to a spring on the exhaust pipe at the ceiling. I would never have guessed it had anything to do with stopping a motor. I pulled hard. Nothing happened. I pulled harder, and the motor stopped. Enormously relieved, I let go. The motor instantly started again. I looked wildly around. Where was John?

He was just coming into the pilothouse. "Pull harder and hold on until the buzzer sounds," he said. "That will tell you that the oil pressure is down, and you can let go then."

I pulled on the rope this time until the motor stopped and a buzzer, a piercing, drill-like tocsin, had replaced the low rumble of the Gardner engine. Then I let go. The buzzer persisted.

"Now push the button that stops that buzzer," John said patiently.

"Where? Where?" I cried again.

"I thought you had seen me do that," he said. "Push that little silver knob on the bulkhead to the right of the steering wheel."

"Where? Where?" was my recurrent cry, especially when instructions were shouted at me in emergencies. Then I forgot the little I knew. It was not until some time during the second year that all the *MoreKelp*'s systems came together for me as if I had been living on her all my life. The rope-and-buzzer routine for shutting off the motor continued to perplex me, because, despite all John's explaining, it remained hazy in my mind. I crawled back behind John into the engine room one day, and he showed me the connections between the big complicated Gardner diesel motor and the mechanical parts above us in the pilothouse, and that helped. I could then visualize what was happening.

At anchor, it was wonderfully quiet. No shrieking wind, no rumbling motor, no buzzers! Quiet. All quiet. John poured himself a Scotch, looked questioningly at me, and, when I held out my mug, poured a smaller one for me. He settled down on the pilothouse seat with his mug and I sat on the stool beside him, nursing mine with both hands.

"Nothing that has to be done on this boat comes naturally to me," I said unhappily. "I don't think I can do any of the things you really need."

"Like what?"

"Oh, cleaning fish, or making up gear, or jumping off the boat and tying it up to docks when you bring it in, and I'm not very good at steering. I don't understand compasses and charts, and I can't do *anything* mechanical."

"You'll learn. You'll learn," he said. He leaned over and took one of my hands in both of his. "All my fishing life, I've dreamed of having someone like you with mental protein between the ears as a permanent partner on my boat. I've fished alone most of my life, so I don't need help there. Besides, with our difference in height, I can't see how you and I both could fish from the stern. The gear is set up too high for you now, and would be too low for me if it was correct for you; and if you stood on a box it would get in the way. No, no, don't worry about any of that! If you cook, and wash our clothes, and tidy up from time to time—because I'm such a messer—and steer when I ask you to and learn to look ahead *all the time* instead of at me, so you can watch out for kelp patches and logs, you'll be a real help. Later on, you might like to ice the fish and turn a hand at making up gear, but the important thing for you to do is to write, and you'll have plenty of time for that. We'll each have our own work. Frankly, I think it's better that way."

He sat back in his seat, and his smile made the small pilothouse sunny, even though it was raining. It was lovely to sit there and watch the pale yellow of the fading sun in the soft rain shimmer along the treetops on the nearby shore and slip away. John said, "I used to dream of having a partner your size who could monkey-wrench that engine in that bloody small engine room that I have to crawl into on my hands and knees. What do you think?" I shook my head, and we both laughed.

"I've had fishing partners—both my boys, principally, whom I took with me one at a time when they were at school, from the age of about eleven, and later when they were working their way through university," John went on. "The eldest one, Dick, used to make wonderful faces out of kelp; he would jab a knife into the globe to make a nose and eyes, and the kelp strings would become hair. He'd put the head over the top of one of the gurdy handles when he was back working with me. The first time, I was a little surprised, but I liked it. It amused me a lot. My younger son, Sean, used to draw, always sketching and painting everything with pen and pencil: sunrises and sunsets, other boats. He was also a good experimental cook; he mixed grapefruit and ham, I remember, and one summer, when it didn't stop raining for more than three hours twice in three months, he did two or three hours of mathematics every day with his back against the exhaust pipe, which is behind the wall now. The boat was very uncomfortable then. It's palatial now, by comparison. As for other partners, I'd rather slave at chores than have to be bad-tempered with youths raised on moron TV and rock radio. When my forty-six-year-old pal from Port Hardy, Peter Spencer, comes for ten days, it's a complete rest for me, because I trust him to take over one hundred percent. All I can do he can do better, and he knows the coast like a book. Most of the modern youth smoke, and I can't stand butts stinking up the place and I worry about fire, which is B-A-D on a boat. The very *worst* is when they talk above the engine, and the younger ones *must* talk. I can only put up with one trip of that stuff. My friend Charlie Walcott, dead now, always took along two deckhands, because they could amuse each other, but you need to make a lot more money to do that and I didn't want to. On the occasions when I have hired a deckhand, however, I had a very simple test for finding out whether he would be a good worker. I found it to be practically infallible. What do you think it was?"

"I can't even guess," I said.

"I gave him a cup of coffee, and if he picked up his mug and put it in the sink when we were done talking, or even better, if he rinsed it out, I would hire him. If he left it wherever we had been sitting, I wouldn't."

John reached up to a small, round, wooden-based barometer that was hanging between two pilothouse windows and casually moved the top needle on its old-fashioned china face from the Gothic print of STORMY to RAIN, and said, "Years ago, I was in Winter Harbour on the west coast of Vancouver Island, tied up to a great big beautiful troller who was making really big money at that time. I just couldn't understand having to support such a huge boat. As long as I could pay all my expenses and educate my kids and save a little money, I just wanted to go on fishing, because I loved it. Another friend told me he wanted to extract a certain number of dollars from the ocean by a certain age and then quit, but I have never had any idea of ending fishing so long as I'm healthy. It's a way of life for me. I just want to take things a little easier as I go along. The people and the sea life of the coast and all its rain and fog and beautiful sunrises and sunsets become part of you and you part of it. It's indivisible. That was quite a squall we went through. On the water, you have to keep in mind all the time that *the sea is out to get you*. Treat the sea as if it is *always* trying to get you. It's the only way to survive. You can never really relax. Never!"

He got up and poured himself another drink and turned around to face me, holding his mug in midair. "Say, do you get seasick?" he asked. "Peter went out with me the trip before you and left his seasick pills for you, in case you do."

"I've never been seasick, but I've never been on a fishing boat, either," I said. "I'm glad those pills are here, just in case."

John resumed his seat. "Am I ever lucky you don't get seasick!" he said. "But if you *do* happen to get seasick, enjoy it. Don't worry over being a nuisance if it happens. It can be very constructive physically, because it cleans out the system. A doctor friend of mine who lives in Crumpet Town—that's Victoria—says I cured his New Zealand nephew of various viruses because I took him trolling and got him so goddam seasick that when the doctor examined him on his return he was well again. The doctor said, 'The ocean completely cleaned him out!' "

"I don't plan to stay if I get seasick," I said.

"Nobody knows how they'll feel in this terrible groundswell until they try it, but chances are if you didn't get sick today you're probably all right," John said. "It's one of the reasons I mostly fish alone. Of the several partners I've had for short periods, the younger they are, the more seasick they seem to get; the tougher they are, the longer they tough it out, and the worse it is for me, because I suffer so for them. You have to break up a twelve-day trip and run some poor seasick man fifteen hours to an airplane."

I had made myself a cup of tea by this time and moved to the bunk to sit against the wall, and John stretched his long legs across the stool I had vacated. "I had this huge husky sixteen-year-old from Prince Rupert with me," he continued. "I had used him in the harbor there to unload fish and wash down the hold: a wonderful worker in the quiet waters at the float. But the poor devil started off the first evening out being seasick, and the following morning, with a northwesterly blowing twenty-five miles an hour and a boat running on either side of us, he lay across the door with his feet in the cabin and his head on the deck, wishing he could die. All I could do was to dash over from the wheel and move him to one side of the door, where I walked across him like a wet sack of spuds for a couple of hours. The next day, he managed to get down a piece of dry toast and sat on the hatch cover swaying back and forth. I was so afraid he might fall overboard in a big swell that the third day, when he didn't improve, I went to quieter waters, into Port Hardy, and put him on a Co-op fish packer and sent him back to Prince Rupert. Peter Spencer was on holiday just then, so he went fishing with me for the next ten days. I had so much grub on board for that seasick kid that Peter *had* to come along to help me eat it! Peter had fished all his life—he was gillnetting briefly with a boat of his own at thirteen—but never commercially after that. We fished that first day until the tide changed and I went to bed. Peter was pulling up the gear and the boat rolled and dumped me out of the bunk. He says I just grabbed the blankets and kept sleeping on the floor. You've already seen how the *MoreKelp* can roll when we're running."

"I wonder whether my cats would have gotten seasick," I said. I had wanted to bring my two cats, Saskia and her son, Angel, with me from New York, but John had vetoed their presence on the boat, so I had found a new home for them back East; but I missed them. In the squall we had just gone through, they would have been terrified. "You were right about those cats of mine," I said.

"If you start a cat on a boat when it's a kitten, it can be quite useful," he said. "I had a Norwegian fishing pal who used his cat as a barometer. Joe was a real tough fisherman, with this very small boat, who trolled a long way offshore. He told me, 'I used to watch my cat, and whenever the cat spat at me and ran up the mast and clawed hell out of it, I knew it was really going to blow and I'd better head for shelter. Then someone gave me a barometer, and every time it dropped a long way down I started running in. Often the wind didn't come, and I lost money. So I got fed up and threw the barometer overboard and said, "Yaw, I go back to the cat." ' "

Late in the afternoon, the rain stopped and the sun came out long enough to create the kind of sunset I had seen on the slide screen in John's kitchen. The dark clouds parted to reveal a red ball that slowly sank in the darkening sky below the horizon, casting silvery and yellow threads on the black ripples of water surrounding us. The boat swung gently at the anchor John had cast over the bow earlier. Ahead were the rocks and trees of the apparently uninhabited shore, an unbroken strip of shadowy darkness; behind us, the open, black, glistening sea.

"Time for supper. Come see where the fresh food is kept, so you'll be able to get it from now on," John said. He bent his head low at the pilothouse door, and I followed him outside to the large box under the boom. The four-by-four-foot lid had a smaller lid in the near corner, which he lifted off and set aside. He reached in and removed a three-by-five clear plastic mat filled with a green, quiltlike material, that he laid on the deck. "This is an ice blanket, made of glass-fiber insulation material," he said. "Very useful for preserving ice."

I knelt and looked in. The hold was filled to the brim with sparkling snowy particles of crystalline ice. Two neat fillets of red salmon were laid out on a board on top of the ice, beside an open white box filled with groceries. A rope attached to it was hooked to the wall at the top of the hatch opening.

"I ice my fish and put them down in the hold as I catch them, starting at the bottom, until I bring them in to sell," John said. "I sold two thousand pounds of salmon at Namu just before you came, and took on three tons of new ice there. I keep all my groceries on top of the ice, except milk and meat, which I bury along the edge." He lifted out the board with the salmon fillets on it and pulled the box up by the rope, set it on deck, and picked out a head of lettuce, new potatoes, broccoli, and fresh peaches; then dropped the box, which contained several egg cartons, cheese, bread, and more fresh vegetables and fruit, back on the ice. He dug the half pint of heavy cream out of the ice, replaced the ice blanket, and put the lid back on the hatch. "Never leave the hatch cover off longer than absolutely necessary," he said. "That's good Namu ice and holds up the best of any I get, but it has to last us ten or twelve days, until we deliver fish again."

Back in the pilothouse, John pulled a seasoned, black cast-iron skillet from one of the shallow shelves under the stove, put it on the hot burner, rolled the fillets in cornmeal, and while he waited for the safflower oil he poured into the skillet to heat, dropped the potatoes into another pot, pumped water into it from the sink, and set the potatoes on the stove to boil. He had already brought down the table across the bunk to get at the salt and pepper behind it, so I set the two plates—his worn tin one and the glass pie plate that became my regular dish—on the nonskid mesh mat inside the table rim. I was beginning to see that every detail on this rocking horse of a boat was designed to utilize every inch of space and to keep everything in place.

The stove interested me. The top, which was not quite two feet wide and a foot and a half deep, with a trough an inch in from the edge to catch fat if it was used as a griddle, was surrounded on four sides by a brass-rod rim an inch or so above it, with vertical fastenings at the corners. A spring hooked from the front to the back rim divided the top in half—a movable barrier that prevented pots from sliding and crashing into one another or onto the floor. There was a black stovepipe at the rear, and right in front of it, embedded in the stove top, was a lever. When you pulled the lever up, it closed an opening in the pipe and sent heat around the minuscule oven instead. The word SEACOOK was stamped in black letters across the white enamel door. John had added color by painting the oven handles a bright marine-paint blue. The stove sat on a little metal apron, and three shallow shelves below held, in addition to the skillet, a saucepan, our red stewpot, a stainless-steel lidded pan with two handles, the old Brown Betty earthenware teapot when it wasn't in use, and two shallow dishpans, one of which served as our bathtub.

The rest of our equipment consisted of a double boiler stowed over the sink, together with a variety of loose items, reasonably secure behind two-inch-high shelf edges. In a storm or a continuing deep swell, the more dangerous movables were stored below the sink behind sliding doors in the cupboard. A fish knife rammed into the crack there kept the doors from popping open and spilling the contents onto the floor, from which, with every deep roll of the boat, they would otherwise have taken off, as if on skids.

John sautéed the salmon until it had a light-brown crust, and at the last minute he dropped dried mint from the garden at home into the pot of boiling potatoes, while I sat on the bunk with my legs curled under me out of his way, and watched him cook. This was different from Christmas, when we had brought prepared food onto the boat in pots from the house. For a refrigerator then we had used the deck corner right outside the pilothouse, in the winter cold. I hadn't even known there was a fish hold, although I had tripped several times over the corner of the large blue hatch lid. After the dirt of New York and the limitations of apartment-house living, the beauty of the rocky coast, thickly covered with trees that extended far into crevasses and over mountaintops that became higher and snowier in the distance, had so dazzled me I could hardly absorb anything else. On my first trip, except for the wall sayings, I had scarcely noticed my immediate surroundings, taking even the stove and its cozy heat for granted.

Now I was alive to everything: the smallness of the boat and the largeness of the rolling sea around us, whose only perimeter was the single black line of coast on one side; the primitive character of the space that confined us—one hundred and forty square feet for working, sleeping, walking, sitting, and eating. I thought, I am floating on the ocean with John in a ten-by-forty-one-foot wooden box that contains sleeping bunks, cooking and storage areas for food, clothing, towels, sheets, and spare engine parts; a hand pump that produces fresh water at the sink from two one-hundred-gallon tanks that I cannot see below deck, where there is also a chugging motor I can hear, fed by fuel tanks also mysteriously hidden whose diesel fumes I can smell; radios, fishing gear, beer, Scotch, wine, even one small split of Mumm's French champagne—but no bathroom. All the other fishing boats I have been on that were the size of the *MoreKelp* have contained a toilet, or head. The only facility on ours was a beat-up galvanized-iron pail, which was flushed out by hand at sea on the end of a long, heavy blue rope tied to its handle and then returned to the deck half-filled with fresh seawater: a movable bathroom that could be set down on the deck facing wherever the view was best, or taken inside the pilothouse. I hadn't even noticed the lack of a head the previous Christmas until we were well out of Pender Harbour. John had nervously brought up the subject as we started up Jervis Inlet. "I am six feet four inches tall, and when I built this new pilothouse I had a choice between putting the floor down far enough so that I could stand up straight when I steer or walk inside, or making room for a head," he had explained. "The one thing I couldn't do was add to the height of the pilothouse, because that would have made the boat top-heavy, and a vessel this narrow can sink in the kind of storms we get on this coast. Not having a head is a protection, too. It's a good excuse for not taking passengers."

I was remembering now the telephone call I had received late one night in New York from John in Pender Harbour a month or so after that first winter visit. It was 2 a.m.—11 p.m. his time—and it began with his usual husky-voiced shout, "Hello!"

"Hello," I said, anxious. "What's wrong?"

"I've just bought a wooden toilet seat that I think will fit very well on top of that pail on the boat," he said. "It's sky-blue, and I paid eight dollars and fifty-six cents for it."

"Lovely," I remember saying. "But it's two o'clock in the morning here. What about it?"

"What about it?" he had shouted back. "Marriage! That's what!"

What a proposal! How could I resist him, and why should I? I laughed myself to sleep.

Frequently over the years, despite the extra height John had given to the pilothouse, he was to complain about the *MoreKelp*'s space limitations. He claimed that his chronic leg cramps were due to lack of walking area, and he jogged regularly in place in the cockpit while he waited for the fish to bite. Two idiosyncratic characteristics of the *MoreKelp* really bothered me. There was no comfortable place to sit, anywhere, and no solid barrier outside on deck between me and the sea. I dreamed about upholstered easy chairs, and carried a square of green foam rubber John cut from a larger piece, which I sat against or on whenever I sat out on deck. A folding beach chair would have done, and we had one on board, but it could only be set up outside when we were at anchor in calm weather.

My real concern was the absence of guardrails on the *MoreKelp*. As a piece of unwanted baggage, I carried a mental picture of myself making a misstep on the slippery deck or losing my balance in rough weather and going over the side. Only a foot-high gunwale separated me out on deck from the water, except in the stern. There the deck dropped several feet behind the fish box where John spent most of his time, making the gunwale hip height for John but chest high for me, and, to my way of thinking, "safe." On the port side, the stern space behind the pilothouse was filled by the dinghy that all commercial fishing vessels are required to have, so I never went there. Ours, pointed toward the sky, hung from a pulley and was lashed to the rigging above the boom. Starboard was the frightening side. I was constantly traveling from pilothouse to stern, carrying food or equipment to John, moving around for exercise, or just being sociable. It was the only cleared walking space we had. John strung up a heavy iron chain along my path at waist height, which was a convenient drying line for dish towels and clothes, but I was always conscious that the yawning space between the chain and the gunwale was large enough for a body (mine) to fall through. I was often asked after that first trip how I had spent my time. "I concentrated on not falling overboard," I said.

Paradoxically, my only serious accident on the boat occurred when it was safely tied at our dock in Pender Harbour in the fall. I was cleaning out the cupboards, and I walked off the boat with a Mason glass jar in each hand and slipped, catching my leg between the boat and the scow, badly twisting my knee, breaking both jars, and cutting my hands. John had given me credit for more common sense than I possessed. His observant troller friend, Reg Payne, at a rendezvous at sea the next season, watching me move around on the *MoreKelp* and then step from our boat to his larger troller, the *Saturnina*, said, "I can see you are not a boat person. A few minutes ago, you told me you were going 'downstairs' to get your jacket. It's *down below*, Edith, *down below*! And by the way, for your own safety, you must always hold on to something solid when you are standing or moving on a boat; never let go with one hand until you are holding on to something else with the other." It was the best piece of boat advice I was ever given.

Except during a fog, when we were trapped in a gray and perilous cocoon, I never felt confined on the *MoreKelp*. Rather, it served as a protection from the outside world; I need only deal with what was immediate and around me. We almost never touched land for ten and sometimes twelve days at a time. When we did, the conversation was invariably the same, whether we were delivering fish at Port Hardy or all the way in to Vancouver; or just plain turning into the harbor to go home.

From me: "Can't we stay out one more day?"

And John: "I know. I feel the same way."

After supper, on that first night of my fishing life, I quickly washed up our dishes and sat down in the seat by the steering wheel to read a *New York Times* I had brought with me. John immediately began preparing his fishing gear for the next day, measuring line for leaders from the big spools hanging from the wall beside his bunk, snapping it off with pliers, tying a release clip to one end, sliding a "lure" on—not a metal spoon like the one he fished with at Christmas but a bright-colored, squidlike plastic object with fluttering inch-long leggy appendages which he referred to as a "hoochy"; and, finally, tying on a tiny swivel and the large, shiny definitive hook.

"Why 'hoochy'?" I asked, as I watched him run his fingers through the boxes of these strange little things. He chose a brilliant cerise one that danced enticingly at the end of the line when he dangled it in the air, and then another, mixed olive-green, yellow, and white, from among the jumble of rainbow-hued hoochies that he had brought up after supper from the bunk under mine.

"We think it resembles a hoochy-koochy dancer with a colored grass skirt," John said. "But in the water it looks like octopus or squid. Squid can change color about twenty-five times in twenty-four hours, so we are constantly changing our hoochies, and that's why there are so many combinations of colors. I especially like this one." He held up the greeny-yellow hoochy. "The springs seem to like them, but if I'm fishing sockeye I use the cerise."

John worked quietly for an hour or so, and I read, until a chugging sound from across the water signaled the arrival of another boat seeking shelter from the wind, like us, in the lee of the shore. It was a smaller troller than ours, and it remained some distance away. I watched a man come out in the bow and drop anchor. John picked up the binoculars and studied the boat for a minute. "I don't recognize the name," he said. "We trollers sometimes know each other better by the names of our boats than by our own names." He reached behind me to an electric panel I hadn't known was there until I turned around. It had five small light switches, marked *Running*, *Anchor*, *Mast*, *Deck*, and *Cabin*; and a sixth, tagged *VHF*.

"Tell me what all these light switches are for," I said.

"*Running* stands for the red port and green starboard side running lights on top of the cabin. You put them on any time you are running from sunset to sunrise," he said. "*Anchor* is for the white light on the end of the boom that shines so brightly it can be seen from any angle; *Mast* is the horizon light at the top of the mast which can be seen all around for three hundred and sixty degrees; *Deck* is an aft floodlight on the cabin roof which illuminates the deck; and *Cabin* is always switched on, so we can turn the lights on in here whenever we want to. *VHF* of course is to switch on that radio."

He threw the switches *Anchor* and *Mast* and sat down heavily on the bunk, slowly taking off his boots. "Jeezus, I'm tired," he said, and an instant later his head sagged on his chest. He was asleep sitting up, still holding one boot in his hand. His head began wobbling and the boot dropped to the floor without arousing him. We rocked slowly in the wake of two more vessels that came in—another troller and a sailboat. I watched out a pilothouse window as the lights at the top of their masts began to twinkle, swaying from side to side in the darkening sky. I tiptoed outside and looked up: we were making our own bright stars in the surrounding darkness with our light at the top of our mast and the one on the boom at the stern.

When I came in, the sound of the door sliding shut woke John up and he looked at his watch. "Ten to eight," he said, reaching up and turning on the light over his bunk. "Time for the CBC news." He took off his other boot, turned up the stove, stirred the contents of a small Pacific Milk can, water, sugar, and cocoa into a pot, and switched on the radio. "God, I'm *really* tired," he said. "I'm glad we're not going anyplace in a hurry. I went to sleep running in 1937, in my second boat, the *Muriel D.* I had had little sleep for two or three weeks of hard fishing, and I decided to take a shortcut through a pass. As soon as I was beyond the rough stuff, I fell asleep. I missed the rocks but piled up on a beach with the boat, sitting straight up. I put the engine in reverse, crawled out on my main pole, and hung all my trolling leads from the end, so that my body weight plus the leads would put the boat on its side and take the keel out of the sand. I was very lucky to have a guy near who put a long rope on me and gave a gentle pull. The boat bounced off in the groundswell, but with an awful bang. Another time, I fell asleep trolling for coho in a flat calm with all my lines out. The boat turned and headed straight for the beach, and I woke up so close I couldn't decide what to do, so I made a sharp left and knocked one pole into the water. To my amazement, my lines were intact and all I lost were a couple of spoons; but it was a terrific shock."

He poured cocoa into our mugs, pulled down the table, and we sat and listened to the eight o'clock news and the weather report that the wind was from the northwest with "sea ripples."

"Sounds good for tomorrow," John said, and snapped off the radio. He rinsed his cup in the sink and drew down a dark green shade attached to the ceiling over the steering wheel, cutting off the view out of the two center wheelhouse windows. He carefully hung his red-and-black-checked wool mackinaw on one exposed wheelhouse window, fastening it underneath with a fishhook, and put a gray Stanfield's over the other, so that all outside light was effectively shut out. I pulled the shades on the two small side windows.

The pilothouse was snug. By the time I had washed my cocoa cup and brushed my teeth, John was sound asleep in his bunk, his glasses resting on the sink counter, the canvas from his bed neatly folded on the steering seat, a red blanket pulled up to the edge of his white beard. I reached above the sink and shut off the bare electric bulb that dimly lit the counter space, and moved cautiously forward in the dark, holding on to familiar objects like the stove and the steering wheel, until

I felt with my right foot the empty space just in front of the folding seat which marked the entrance to my quarters below. Sitting down on the floor, I reached with my right leg until my shoe rested on a bilge stringer, a timber running along the inside of the hull, and slowly let myself down into the fo'c'sle. When both feet were solidly planted on the board floor, I climbed up on the edge of the lower bunk and groped around above the upper one until I found my bunk light and turned it on.

After John's sons grew up and moved away, he sometimes took the son of Jimmy Reid, the local boatbuilder and fisherman, as his deck-hand, or an occasional passenger who wanted to sample commercial fishing to see whether he should take a fling at it; and whenever he could come, Peter Spencer, who had left me the seasick pills. Peter had taken his vacation from his jobs as oil deliveryman and volunteer fire chief of Port Hardy to fish with John for those ten days before my arrival. "He's buying a boat now and will be fishing for himself," John said. Peter was the first fisherman I ever heard of who got seasick, but he never mentioned it and was matter-of-fact about taking pills. I subsequently found out that there are quite a few seasick fishermen. I would have needed no persuasion to stay ashore had I gotten seasick—not after I experienced the steady ocean groundswells that produce days of endless boat motion.

"For God's sake, can't you stop rocking even for a moment and stand still?" I shouted one morning at the *MoreKelp*, while I was trying to pare and slice vegetables for a fish stew, braced against the sink. I was exhausted from the effort required to stand up and keep everything from sliding to the floor.

"Good, good," John exclaimed, with a kind of delight, coming in for morning coffee at that moment. "You'll make an excellent fishwife."

These recent occupants of the upper bow bunk had slept at the bow stem facing the stern, with a cheery reading lamp on the wall over the bunk. I didn't like riding backwards when the boat was moving, and, even with the comfort of the reading lamp in that windowless corner, I felt claustrophobic for the first time in my life. After thirty minutes of this, I turned around so that I was facing forward, with my feet at the bow instead and my head farther back under the fo'c'sle hatch cover, which had a round glass port that served as my skylight.

Any change from the familiar routine on a small boat has to be thought out carefully. In my bunk, it was head space; I could never again sit up in bed on the *MoreKelp*. My pillow kept my head higher than my feet, but, because of the downward slant from the bow of the deck above my head, I had only fourteen inches now between my bunk and the beams above. John ran an extension cord along the wall from the bow socket to give me light, and I read or drank my morning tea slumped down on the pillow or on my side, to avoid bumping my head on the underside of the bow deck. I had to roll in and out of bed; but I had open air on my face, a fresh breeze and stars above me on a clear night through my skylight—my private universe.

Ever after, each night before I went to sleep, I stood up on the edge of the lower bunk with my head inside the hatch cover and raised the heavy lid to insert a small, six-ounce can of evaporated Pacific Milk between the hatch cover and the deck to make an opening in my skylight for the sea breeze to come in. I experimented that first night with several can sizes: a flat sardine can opened it a mere crack; a Campbell's soup tin pitched the hatch lid at too wide an angle and let in the drizzling rain from a sudden squall; but the milk can provided a perfect slanted opening and support for the hefty raised lid.

Once I had rolled into my bunk, it was very comfortable—far better than some of the miserable sleeping slots below deck I have seen on many older boats. John had painted the fo'c'sle a dazzling white with "good marine paint" to celebrate my arrival. His dedication to glossy marine paint was absolute, and whatever was left over from painting the boat appeared on the walls of the Garden Bay house. We had grass green in our bedroom, battleship gray in the hall, and the sick bottle green in the bathroom. Leftover white marine paint with a touch of blue to make it whiter eventually transformed our house. I had to make many trips East, because my mother, who was in her eighties, was ill. Each time I returned to Garden Bay, some surprise around the house awaited me. Once, it was a spice shelf; another time, a wire cage around my Queen Elizabeth rose to keep the deer out; and, after a particularly long absence, a white back wall in the living room, painted over dark-brown, wafer-thin airplane-spruce paneling, and a dazzling white kitchen, bathroom, and hall. Once started, John never stopped until he ran out of paint.

V

The morning after my arrival on the *MoreKelp*, I awoke in the dark to the realization that the lapping of the waves against the boat's hull and the mild roll at anchor had grown into a roar of rushing water through which our vessel was moving. The sky was black above my overhead window, so I turned on the light and looked at my watch. It was 4:30 a.m. I lay in the bunk, which was dipping up and down, and listened to the thrum of the engine, at the same level belowdeck, behind a partition at my head.

On my side, the partition was painted white and festooned with chains, pulleys, and little wheels going back and forth, up and down; electrical wiring in black, white, green, yellow, or red casing stapled or bracketed to the wallboard; and cerise lures, with black polka-dot spots, that made a light tinkling sound swinging from the small hooks. I turned over carefully, with my head propped on my elbows, and was examining this attractive display when John appeared with a mug of tea. "I don't know what this panel is, but you'd win a prize with it in an art show. You could attach a little electric motor and keep everything moving," I said.

"It's part of the boat's steering mechanism, the system of cogs and wheels, pulleys and chains that make the rudder respond when the steering wheel is moved by hand or automatically," John said. "I like to keep everything in the open that might need monkey-wrenching, so I can get at it in a hurry. The pulleys you see moving in a cyclical motion now are connected to the automatic pilot that is steering the boat while I am down here to bring you morning tea and take a look at the engine. The colored wires are part of the electrical system—for the lights, the automatic pilot, depth sounders, radios, generator, and so on. I can see at a glance what's what from the colors."

He picked the spoons off the hooks and put them in his pocket. "I hung these here to dry by the engine after I painted them and forgot where I had put them." He laughed. "Age," he said, and tapped his head.

He pulled a cord fastened to the edge of a board in back of my head which turned on the light of the engine room, dropped to his hands and knees, and crawled down a narrow passageway beside the Gardner motor, where I could hear him groaning and swearing; then he was quiet, and I pictured him fiddling with the engine. On his way back, he picked up my empty mug, which I had tucked just inside the rim of the bunk below. "God, what a bloody uncomfortable spot that engine room is for someone as tall as I am!" he exclaimed. "That Gardner diesel is the best trolling engine there is. With its transmission, it weighs twenty-five hundred pounds, has five cylinders, seventy-two horsepower, and a trolling speed of one and a half knots. Perfect. But I hate crawling around on my hands and knees." He looked at me wistfully. "It would be so easy for you."

"I can't even hammer a nail," I said.

He patted my head. "Don't get up for a while. I've been fishing alone so long that I'm used to having the place to myself first thing in the morning, so I can plan my day. Unless there's a beautiful sunrise, of course."

He stepped back up into the pilothouse and I watched the wheels and pulleys spin around as he took the boat off the automatic pilot and steered by hand, to a sharper rhythm. He opened the throttle, and the *MoreKelp* leapt forward, hurrying to the fishing grounds.

I lay in my bunk, glad to stay there until the fo'c'sle warmed up now that we were under way. The stove kept the pilothouse warm, but the Gardner engine was the only source of heat below. The night before, it had been turned off the minute we dropped anchor, and by morning a good damp chill had built up around my bunk. Almost all our fishing was done three or four hundred miles north of our home, halfway up the British Columbia coast, officially referred to as the Central Fishing Area. During four seasons I spent with John on the *MoreKelp*, there were only a few summer days when I could walk around without wearing my red wool turtleneck sweater, and I usually added my gray Stanfield's when I went out on deck, or my "floater," a quilted nylon jacket designed to keep afloat a body that goes overboard.

From my nest of flannel sheets and wool blankets, I had a view—which always gave me pleasure—of glossy white broad structural boards forming the sheer line in a graceful curve up to the bow stem, and of the lattice of cross and long deck beams overhead. Two spare gaffs hanging from the boards opposite, glossy white with handles wrapped, one with yellow tape, the other with crimson, made vivid dashes of color above the long, white bench—narrowing to a point at the bow—that ran along the starboard side.

A shout from John: the word "sunrise" sounded above the motor's rumble. Shivering, I rolled out of bed to the small triangle of floor space between my bunk and the bench, where I had left my clothes piled up, one set inside the other. Like all fishermen, John never completely undressed; so he could be ready for any emergency. I had learned on the trip to Jervis Inlet to dress with only one instant of exposure, when I pulled my red flannel nightgown over my head and inserted my head and arms through all my upper day garments in a single motion. The boat had been at anchor then. Now, on the running vessel, I danced a brief balancing act when I stood up to pull on my bluejeans; a lurch of the boat sent me abruptly back to a sitting position on the bench, to pull off my nightgown, replace it with the prepared top set of garments, and put on my socks and sneakers.

Fully dressed, I climbed onto the edge of the lower bunk long enough to stand up with my head inside the hatch cover, remove the milk can, and drop the lid into place. I made my bed, tucking my nightgown under the pillow and putting a clear plastic cover John had given me over the whole bunk, with the milk can in the middle to hold it down. Later, in rough weather, when waves broke over the bow, seawater and rain from the torrential downpours leaked through at the edges of the hatch cover, leaving a large puddle in the center of the plastic. I had to rearrange the plastic at a slant, so that the water that leaked in could flow over the bed to the floor, where I could mop it up easily with paper towels. I put the book I had been reading in a net bag stretched along the port wall by my bunk, and threw my toilet bag, glasses, a bag of unread newspapers and books, needlepoint, and the notebook and pencils I always kept within reach up on the pilothouse floor. Then, in a movement that became routine, using a boat stringer as a step, I shoved my body from below through the open space into the pilothouse, pulling myself to my knees on the cabin floor, and scrambled to a standing position. If John happened to be steering, as he was then, I used his right leg as an anchor to pull myself up; otherwise, I grabbed any convenient post.

"Why don't you put steps there?" I asked crossly that first morning.

"My legs are twice as long as yours, so I didn't need any," he replied.

Later, when I went down below to examine the store of canned goods under the fo'c'sle bench seat, a wooden box had been set on the floor upside down as a step for me. I could see how much in the way it was, so I put it back in the engine room, where I had last seen it, and said to myself, You must stop complaining about things you don't understand.

John pulled the stool down while I was brushing my teeth at the sink in the dark, and I came and sat beside him, grateful for the mug of hot coffee he handed me. "Toast on the back of the stove," he shouted over the motor's noisy rumble. "We're going to have a glorious sunrise. I didn't want you to miss it."

I could only see that it was still night; the otherwise black sky was illuminated by a full moon, whose brilliant light cast a shining path down which our vessel seemed to be traveling, our bow raising a phosphorescent foam as it cut through the dark water. John touched my arm and pointed: a hint of orange appeared at the line between sea and sky, followed by a burst of orange-red to herald the rise of the flaming oval sun in the heavens, sending out streaks of fire that lit up the whole world through which we, only we, were moving. John turned off the motor and we drifted, gliding silently on the rippling sea. "The choice in views for beauty along this British Columbia coast completely overwhelms me," he said. "I love it so deeply that I want to put my arms around all of it. I want to pour the whole coast over myself!"

For one instant, I turned around to pick up my toast from the stove, and when I turned back it was light outside; the day had begun. John pressed the shiny chrome starter button for the motor, pushed the throttle forward, and we were moving rapidly again. He leaned back against the side window and pulled down the chart board from the ceiling. It fell across the wheel, spilling charts into his lap. He placed them back in the rack, pulled one out, putting it on top of the others, and examined it while he continued to steer. He drew me over beside him so that I could see as he traced the path of our voyage: from our overnight anchorage just to the west of Fitz Hugh Sound, through a channel marked Hakai Passage, into the open Pacific. The compass needle pointed directly west.

"If we kept going straight ahead, where would we land?" I asked.

"Japan, China," he said. "B.C. is the end. I always get a thrill to think that there is nought left to the west between me and Japan and China but ocean for three or four thousand miles; and row upon row of snowcapped mainland mountains as far as I can see to the east. The Gulf area where we live is pretty, but it's dead, dead, by comparison with the life movement up here: blue ocean rolling gently; the vastness of a mountain sky."

We were at the end of Hakai Passage. He turned the wheel to the right, and the compass swung from W to WNW. Land, to the starboard, consisted of islands separated by indentations—sounds leading into channels that carved deep fjords into the mainland coast. John made a circle with his finger on the chart around the outermost land group: a narrow little island with three tiny islets below its south end, all four surrounded by a border of shallows, colored blue on the chart, whose depth readings were all less than ten fathoms. "We're going to Goose Island and the Goslings," he said. "I've been fishing this area steadily since 1949."

John smoothed the charts down and fastened the board back inside its ceiling hinge. "You steer. I'm going to put the gear out now," he said. "Take a compass course of north-northwest, and keep checking up on yourself by looking at the chart. Read your charts properly and you'll know your direction and depths. Watch for floating logs and seaweed! Big clumps of kelp wreck the gear and a log can sink us. Like that one." He pointed to a harmless-looking log end about two feet in diameter bobbing up and down in the water ahead of us. He turned the boat so that we made a wide detour around it, and then said, "A deadhead like that is one of the most dangerous hazards of the sea. You are looking at the end surface, but there could be eighty to a hundred foot of log standing straight up under that water, and if we hit that it could make a hole in our wooden hull big enough to sink us; if we rode over it, we might disappear without leaving a trace. *Watch out for deadheads*!"

I stood up, so that he could get past me, and as I took the wheel he flicked a switch on the side of a foot-square box hanging at eye level from the starboard wall. Its round face resembled a clock, with "o" in the twelve o'clock position and markings at intervals of ten, up to two hundred. A dot-sized, flashing orange light was circling and had stopped, pulsating, at 80.

"We have eighty fathoms of water beneath us now," John said. "We call this machine a flasher-sounder. It takes soundings at the bottom of the sea and tells you in fathoms how deep the water is below the boat." The orange light was circling again, then stopped briefly at 23. John instantly turned the wheel, and the light ran ahead to 82. "We just passed over a jagged rock or reef when the sounder hit 23," John said. "If I had had my gear out, it might have ripped off. You have to move quickly. Steer to forty fathoms." He was putting on his plaid mackinaw as he talked and he poured himself coffee from the open saucepan on the stove. He leaned over, mug in hand, and turned on the big radio to the CBC news. He switched the sound outside, and I could hear the CBC announcer's voice giving the six o'clock news from the loudspeaker in the stern. He departed, shutting the pilothouse door behind him.

John was asleep the first time I steered the boat, when we were in Jervis Inlet. This time, he's awake and will see everything I do, I thought; I'd better look at the chart and find out where I am. I stood up on the seat, holding on to the wheel, and let down the chart rack, which descended with a crash, spilling charts over the compass shelf, the seat, and the floor. I tried to pick them up, tried to steer, tried to look at the compass and the sounder, all at the same time. The *MoreKelp*, with a mind of its own, turned sharply to port: W on the compass. A firm turn of the wheel by an Unseen Hand (John's) and we were back on course again. I laughed out loud, and settled down to read the chart, holding it with my elbow this time while I glanced at the compass, at the sea ahead, and at the land alongside us, checking in all directions at once, keeping a finger on the wheel—my message to the Unseen Hand that I was also there. I never was able to bring the charts down with any grace while we were moving. No matter what I did, I could not halt their slippery descent onto the seat, my lap, and the floor. They were different sizes, but most measured close to three by four feet; I held the one I wanted with my teeth until I could flatten it down and put it behind me, temporarily, on the seat.

I gave up steering and hitched the chart rack up securely again, using both hands; put on my Stanfield's over my shirt and sweater; poured myself a cup of hot tea, another for John; and put two pieces of whole-wheat bread on a little wire rack I placed over the hot spot on the stove to make toast. While I was turning the slices, I watched the wheel swing firmly to port, a little farther to port, then make a quick starboard turn, guided by that Unseen Hand. With four steering stations on the boat, John was always within reach of a wheel. In addition to this main steering mechanism in the pilothouse, there was the small, six-spoked galvanized-iron ring with a green center, a foot in diameter, outside on the front wall of the pilothouse below the windows; another two-spoked red wheel on the back of the pilothouse, with which he steered while he was working on the main deck; and a big aluminum-painted steering wheel, taken from an old car, in the cockpit, with a plywood star lashed across which made it easier to grip in wet weather with slimy gloves.

When the toast was done, I added margarine and John's favorite Seville orange marmalade, made annually for him in exchange for fish by a logger friend, Dick Donovan. Holding the two mugs full of tea by their handles in one hand and the toast in the other, I went outside, walking carefully. The mainland had receded, and I turned for a second to look ahead, where I could just make out in the gray illumination of the new day a chain of rocky islets, on the first of which a light blinked an off-and-on warning. Beyond was a much larger wooded mass, and then, nothing! A last outpost, I thought, before Japan. I delivered John's tea to him, and when I was safely settled with my mug on the deck, my back against the gunwale and facing John on the other side of the fish box, I said, "I always assumed it would be quiet on a boat. But it's so noisy here! I call the noises the Buzzer People. To begin with, there's Old Anchor—a real groaner."

"That's right," John said. "That's a Danforth anchor, the best kind for mud or sand. It lays flat, so the boat can change direction without snagging or pulling the anchor out. It weighs seventy-six pounds, and last night it went down, groaning, twelve fathoms. I paid a dollar a pound when I bought it, and now it would cost a couple of hundred dollars. I have a forty-pound spare on the boat, and another one at home. I don't know where you got the idea that boats are quiet. The teakettle is what drives me crazy. That bloody rattling! I talk to it all the time. I tell it to quiet down, staggering all over the stove the way it does! When you're traveling, the vibration from the engine makes it walk. 'Stop that rattling!' I shout. 'You mustn't do that!' and I swear at it, bang it, and generally end up taking the lid off. Then there's the sink. It was gurgling this morning, and I said, 'I can't stand you making that noise! I'm going to ram you with a rag!' Which I did. You should have been here before I got my hydraulic system. I had a set of V-belts to drive the gurdies which sounded like a chorus of amateur Wagnerian sopranos, every time I trod the pedal that engaged them."

"The Chief Buzzer is that piercing voice—the long, horrible bzzzzz I hear when I go through that confusing routine with the green rope to turn off the motor," I said.

"But it's so simple!" John exclaimed. "Your trouble is that you don't hold on to the rope that cuts off the fuel supply long enough for the fuel to stop burning, so the engine starts up again. All I have to do after I've anchored or tied up at some dock is to pull on that rope as I come in the door. What could be more convenient?"

"I'm going to try to remember," I said. "In my mind, that green rope is a mute Buzzer, because it stops the wungety wungety wungety of the motor. Then there's that deep throaty gurgle of the engine's starter button, beside the pilothouse steering wheel, and the racket the heavy iron chain makes that's attached to the anchor. When you push the clutch for the anchor winch with your foot outside the pilothouse door to get the anchor moving, it wheezes and grumbles; then rattle, rattle, rattle, while you run around to kick the chain along, and r-rmp!—the shackle goes over the bow roller, and wz-z—the rope travels along the deck. And when all that chain has clattered over the side, and the rope it is attached to rolls off the winch, have you noticed? It gives a little sigh, sssh-sssh, as it hits bottom!"

"Those are fine noises. I can see you've been giving this matter your full attention," John said. "Boats are full of creaks and other odd sounds. There was what Dick used to call the Great Mysterious Noise, when we were fishing off the west coast of Vancouver Island. We'd come in, drop anchor, shut off the engine, and the only sound would be the boat rocking, until there'd be this new noise—shoosh shoosh shoosh—we couldn't identify. I'd be up the mast and down in the engine room, peering into the fuel tanks with my flashlight, getting all worked up but never finding out what it was. Then it would just stop." He was washing down a fish. "And what about this hydraulic pump?" he asked. "Listen to it howl!"

After a pause, he went on, "There are two more Buzzer People you haven't heard yet. They ring bells. There's the heat sensor in that round white disk in the ceiling above the sink that gives off a piercing clang, like a fire alarm, if the pilothouse gets too hot. The other is the bilge alarm. If the boat leaks and water rises in the bilge to the level where the alarm is attached, it screams loud enough to wake the dead. I'm proud to say that both these safety regulations were put in while I was one of the directors of our Pacific Coast Fishermen's Mutual Marine Insurance Company, which we fishermen organized for ourselves. Speaking of fire alarms, I've installed a new extinguisher in the bow for you. If we have a fire, remember to get yourself out first and *then* use the fire extinguisher! There are three aboard: in the pilothouse by the door, on the fo'c'sle bulkhead as you go down, and outside on the deck. Most fishermen and yachtsmen have only one, but I believe in the maximum."

Before I came out, John had dropped down the poles from their vertical running position of the previous day and he was putting the gear on the spread-out lines while he talked. His hands, encased in heavy black rubber gloves, moved rhythmically back and forth over the lines, one by one, taking out the little boards that separated the set of leaders for each line, which he always inserted whenever he put all the gear away, last thing in the day, or brought in all his lines to drift and take a nap. He lifted the long leaders that hung, partially coiled, fastened to a little wire on the cockpit gunwale, with hooks and lures dangling from them, out of the pail in the stern where they rested in a bright yellow fluid that kept them from rusting, and clipped them at regular intervals to the lines, stopping briefly to file the points on the hooks before dropping them into the sea.

"It takes me about fifteen minutes to get out the gear, which I always leave ready to go the night before," he said. "The mainlines of twisted stainless-steel wire are six-hundred-pound test, made specially for trolling, and are a straight pulley arrangement on the poles." He put his hand on the brake of one of the bronze spools that were mounted on heavy planks along either side of the fish box. "When you came out with me last winter, I was sportfishing and was allowed one hand-driven gurdy, but now we're fishing in the commercial season, and these gurdies are power-driven winches, with hydraulic levers that pull the lines in and run them out again, with a slow braking release. We need that power because each of my six lines is weighted down in the water with one of those leads you saw—anywhere from twenty to sixty pounds on my deepest lines, depending on its position."

"What a funny name: gurdy," I said.

"Somebody just called it a hurdy-gurdy, and it stuck," John said. "It was named for the merry-go-round—something that goes around driven by power. In the nineteen-twenties, guys started trolling with rowboats, handlining, but in the early thirties they began using single-cylinder gas engines in their rowboats, which gave them the power for the gurdy. I started to fish then, and out of thirty boats only two had power-driven gurdies, which didn't work very well. I got my first ones in 1936, on consignment from a machine-shop guy, Jack Swann, one of the best machine mechanics the fishing industry ever had. He gave us unlimited credit and asked us to try out his machines and tell him about them, and then he would improve them. They had very poor brakes, which would either stop suddenly or not at all, so you broke off a lot of gear. Mine worked every once in a while; I almost always ended up having to pull the lines in by hand. Now everybody's got hydraulic gurdies, and they work very well. There's fifty to a hundred fathoms of trolling wire attached to each lead cannonball running from the spool around the gurdy, on each line."

I was looking at his fishing poles, like so many things on the boat, as if I had never seen them before. There were four: two pointing toward the sky at a forty-five-degree angle from either side off the gunwales just ahead of my bow hatch opening; two behind the pilot-house, spreading directly outward from their heavy brass hinges fastened to the deck, long arms reaching seaward over the starboard and port gunwales.

"I love it when the poles are down," I said. "When we look like some exceedingly large winged thing floating on the water."

"I'm trying to decide whether or not to take the bow poles off altogether next year," John said. "The *MoreKelp* is a lot more top-heavy with them on, and when I'm bucking bad weather I usually have to pull them up because the leads are liable to bust your gear, bouncing. The boat would be more stable without them, I think, less likely to roll without that top weight, and it would be less work for me. The fishermen who've taken them off say they get just as many fish on four lines as six; that means only four gurdies instead of six."

"How can you have six lines with only four poles; and four lines if you have only two poles?" I asked.

"First thing to know is that the three lines on each side must drag along at different positions to the back so they don't tangle," he replied. "The bow pole has one line dropping from it right into the water with the heaviest lead on it—fifty pounds—to keep it free from the others. The two main poles have two lines each: the mainline hangs from the top of the pole with about forty pounds of lead on it straight down into the water; and a second line that swings clear of the boat by means of special rigging consisting of snaps, stoppers, and a plastic float, called a "pig," with a fifteen-to-twenty-pound lead on it. We call that the "pig line." He held up two little colored-plastic doughnuts, one smaller than the other. "When these stoppers on the rigging meet, the pig line swings out and is carried away from the boat off the stern by means of a Styrofoam float that I attach as I run out the line. We used to use pennies for stoppers—with a hole drilled through the middle—instead of these doughnuts, because they were cheaper."

He reached over into the farthest section of the fish box, where two large neat white Styrofoam rectangles in red plastic frames were stacked, and lifted one up. "When I clamp this on the line, it makes it spread out at a higher level and keeps it clear of the propeller and other lines. This block of Styrofoam has to be large to float the weight of the lead and line. I don't know exactly why we call it a 'pig,' but the earliest I saw were made of canvas and automobile inner tubes that looked to us like pigs," he said, dropping the Styrofoam pig back in the fish box on top of its twin. "They don't now, but they used to."

"I'm trying not to be confused," I said. "Let me tell you the way that basic arrangement looks to me and see if I have it right. I feel the first important thing for me to remember is that the short lines hanging into the water are what you call the leaders, and the hooks at their ends are what catch the fish."

He gave me a strange look. "That's right," he said gently.

I took a deep breath. "And those short lines made of the clear plastic that you call monofilament are clipped on, one by one, so many fathoms apart, to the mainline that is spooled on the gurdy. When the spool turns, the mainline runs through those springs and various attachments on the poles, out from the boat."

"Yes." He ran his finger uneasily around the nape of his neck.

"And those mainlines are what carry the small lines out from the boat and down into the water," I persisted, making a full circle in my thinking.

"Right!"

"And when the little springs attached from the very tops of the poles to the mainlines start to quiver, instead of stretching back and forth with the rhythm of the boat as it moves through the water, that's when you, and even I, can tell there's a bite."

"Correct!"

Neither of us was fooled by my superficial grasp of the gear. The first few weeks especially, I kept looking over the side to see if John had caught a fish and, despite the complicated mechanisms he employed—or maybe because of them—I was constantly reminded of a game I played when I was not more than five years old, called Go Fish. A supply of shiny metal fish was hidden on the floor behind an upright cardboard screen painted like the sea. The idea was to cast the magnet over the frame and catch as many fish as you could. That simple game flashed across my mind for years, while I was trying to puzzle out John's operation, and I am not sure John ever thought I did. I remained blissfully baffled, but I felt I tried, and he appeared to remain blissfully content to have me there and trying.

John was putting out gear on the starboard bowline while I talked. He shifted his hand to the second gurdy, which controlled the mainline, and threw over the side a long leader already attached to the line—the transparent monofilament with a lure and hook at the end. Then he braked the line, and, after it stopped, carefully lifted a lead ball from its rounded resting frame beside the gurdies and let it drop over the gunwale. It made a dull, thudding noise on its way down, knocking against a piece of rubber belting nailed to the outside planks of the boat to protect the wooden hull.

"That's my heaviest lead cannonball—fifty pounds—on this bowline," he said. "The first leader is attached right above the cannonball and I always toss it into the chuck first, before I lower the lead, as a safety measure. If that heavy lead by some mischance were to run away from me on its way down while I was still holding the lure and hook, it could rip my hand apart!"

With his left hand operating the hydraulics handle, John went on moving the line from the gurdy spool out to the pole, running the twisted steel line through the fingers of his right hand until he came to two quarter-inch brass markers about six inches apart. He stopped the line, snapped another leader onto the wire between them, then pushed the gurdy handle forward to let more line out again. He explained, "The marker clips for the leaders are set on the bowlines every two and a half fathoms—three fathoms on the main and pig lines. In the springtime, fishing spring salmon, I use less gear. The total? Oh, I'd say there are thirty-six to forty hooks in the water altogether, on all the lines, when all the gear is out. Ideally, I should haul in all the gear three times an hour, because you pick up a lot of junk like sole, cod, or hake, or jellyfish, which corrode the hooks; and if anything is hanging on a line you won't catch a salmon. I guess I average three times every two hours, generally starting with the bowlines. Coho seem to bite more on my bow poles, maybe because they have a springy action. I go around all six lines, bring in what's on the gear, immediately kill and clean the fish I catch, and put them in the checkers—that's what we call these divisions in the fish box that check the fish so they won't slide all over the place."

In a couple of moments, he continued, "I leave them there, covered with burlap sacks for one and a half to two hours. Then I rewash them, because, although I have cleaned out all the blood right away, the fish continue to bleed. I want to be sure I get all the blood out before I put the fish down to cool on the ice for half an hour. Then I go down the hatch into the hold and place the salmon in rows on their backs, tail to tail, and poke a bit of crushed ice into their bellies. When I have put down a whole layer of fish, I place a wooden pen board in an insert that is there to keep them in place. I sprinkle them lightly with more ice and start a new layer. You have to watch everything, including the ends of the fish box. Ice tends to melt away from wood, so you must never let the ice melt away from the edge of the hull in the hold."

He had thrown the first leader and hook over the side, and dropped the cannonball behind it. "If you are careful, you can keep coho and spring salmon up to ten to twelve days in tiptop condition," he said. "The more hours you put in fishing from daylight to dark, the more fish you catch, if you can stand it. We'll be getting up generally from four-thirty to five o'clock every morning, and then we'll run for anywhere from forty minutes to two hours from where we anchored overnight to the fishing grounds. Then we just troll until dark." He put his rubber-gloved hand to his throat. "I'm talking too much. It isn't good for me. I have a weak voice, like my dad." He moved his hand to the handle of the third and last gurdy on the starboard side, pushed it forward, and the wire began to unwind from the spool. "Watch now. I'm going to set out the pig line off the stern."

I looked up. A set of doughnut stoppers on the line caught a connecting device attached to the rigging on the main pole, and I saw a line that had been flapping in the wind at the pole swing out and away from the boat behind us, creating a graceful arc from pole to water. When he had let out the line partway, he took one of the two pigs from the fish box and snapped it horizontally on the line, using a clamp that was part of the rectangle's red frame. He released the brake and watched the pig zoom down to the sea, veer away from the boat, and course through the water behind us. He moved over to the port gurdies, and the starboard pig was shortly joined by its partner on the port side: two friendly white blocks, each with a corner exposed so as to form two white triangles above-water, which followed faithfully at a steady pace two or three boat lengths behind us.

On sunny days, when I was steering straight ahead, I loved to lean far out the bow window to look back at the pigs—a comforting sight. In bad weather or when I was executing a turn, I *had* to keep looking back to make sure the lines had not become entangled. The angle of the turn, the force of the wind, or the tide—any or all three of these could provoke that disaster, and cause John to lose time, or worse, the gear if he had to bring in the lines and straighten them out.

John dropped another cannonball over the side and picked another leader from the pail. He continued uncoiling leaders and lures, one after the other, methodically tossing one end into the sea and clipping the other end on the steel mainline—a routine I was to witness dozens of times every day and day after day, as long as we fished.

"It looks as if I'm doing the same thing over and over, but I've got a lot of things to think about at one time," he said. "I may be sharpening hooks, thinking about where to fish tomorrow, planning to change the lures as the color of the water changes, rearranging the depths of the lines to match the changing set of the tides, watching the poles to see if I have caught a fish, and keeping an eye out for beds of knotted kelp and half-submerged logs. If it's misty, I'd better be watching for boats, too, coming suddenly out of the haze. If it's foggy, I'll be listening for the breakers on shore to tell me I'm too close in, and before you came on board, my dear, I might be cooking a stew as well. Meanwhile, I am hauling in and letting out lines. They won't wait; and I am reading the clouds and listening to the weather reports, as well as to all the stories over the radiophones on what is being said and *not* being said about where the fish are. All this time, I am also bearing in mind the state of the fuel and water supplies, and where I was last year and where I caught fish in past relevant fish cycles. The one thing I *don't* have to worry about is paying off a quarter-of-a-million-dollar mortgage on the boat the way some of these young guys do, because it was paid for long ago."

He was about to clip a leader to the line when he stopped and held it up first, for me to see what he had put on it to attract the fish. At the end of the gut was the cerise-colored, spoon-shaped metal lure, and below that the blue-black hook. "We call this Lupac spoon an Easter Egg," he said. "I don't know why, but Easter Egg spoons are very effective—especially later on near the spawning season. I'm a great believer in this stuff for murky waters, and in very clear water I like murky colors." He clipped the leader on the line and pushed the gurdy handle forward. The line spun out away from the boat and arched like a bowstring to the point where it entered the sea.

He picked up another leader. This one had a narrow foot-long rectangular chrome-and-red metal plate attached to a swivel at the end of the gut, followed by another swivel, a short length of gut line, his favorite yellow-green hoochy, and the inevitable black Norwegian Mustad hook. He bent the plate slightly with both hands. "We call this a flasher. It's another kind of lure to attract fish," he said. "That bend gives it my own special wobble." He ran the leader out, and I leaned over to watch it drop behind us into the water, the flasher creating the illusion of a silvery fish, twisting and turning just below the surface before it disappeared in our wake. He added several more leaders with hoochies or spoons on them, then picked up a small gray-painted six-inch-long plastic fish with yellow saucer eyes with round black pupils, a red open mouth, and a pointed tail. Using pliers, he removed the metal spoon that had been on that leader and attached the little fish—which looked to me like a charming toy—in its place, adding a swivel, then a hook. "This is a plug—another kind of lure," he said. "Plugs come in all colors. We use them when there are herring around, mostly, because they resemble the herring the salmon feed on. They have a kind of wiggling action behind the boat. They worked well in 1956, the first year I used them, and I didn't know what they were supposed to do. Now I order a half dozen every spring when I get new gear. I don't know whether it's their speed in the water or a sexy wiggle, but sometimes they catch fish, so we'll put one on. I've used all these lures before, but if I'm trying a new one I don't put it in a favored place like the bottom line. I try it in a bad place. Then, if it works, I put more on. Fish are like children at a party—some youngsters like chocolate cake, and some like another kind better."

All the gear was soon out, trim and neat, the lines plunging taut from the outspread poles into the water, carrying the leaders, hidden now, into that chilly wet underworld; and at a respectful distance behind our boat were the pigs, cutting two graceful, sweeping strips through spray-bathed pathways sixty or more feet apart.

We had slowed down to two knots. I stood up and stretched, observing the lines of the *MoreKelp*: narrow beam, tall poles, high mast, with its thin ladder up to the crosstree. I looked at John, who had leaned down to twist the wheel and change our course away from shore, out to sea. "You have an elegant ship," I said.

John straightened up and slowly began pulling off his heavy gloves.

"Do you always wear gloves?" I asked.

"I never touch my gear without my gloves on," John replied. "Fish are repelled by the smell of the human hand in particular, but the scents of sea-lion flippers, tobacco, lubricating or diesel oil are just as bad. Occasionally, when I wasn't wearing gloves, I've gotten oil from my hands on the lines or leaders, and that particular line stopped catching fish for quite a while. Some of us trollers think scent is so important that we have even rubbed our spoons in fish blood or roe in the belief that this will attract fish."

He was sitting now at ease, on the edge of the stern rail, one foot propped against the fish box, holding on to the knotted line tied to the boom. A small plane droned overhead, and he looked up and waved. The plane dipped its wing in response and kept on going. "That makes me think of my first experience selling fish to an airplane," John said.

I took my notebook from my pocket and, as soon as I had settled on the edge of the fish box, he began his account. "It was off Stillwater, not far from home, three or four years ago. It was a beautiful sunny day and I was trolling, half asleep, when I heard this buzzing and saw a plane coming in the distance over the top of Texada Island. I didn't think very much about it and then it headed straight for me. My first reaction was, 'What have I done wrong?' and 'It must be somebody from the Fisheries Department coming to inspect me.' Then the plane came down and started circling around me and I could see it was a private charter plane. It landed right off the end of my poles and circled around again. Then the pilot jumped out on the pontoon and shouted at me, 'Will you sell any fish?' I was most surprised at this, and it took him one more circle before I could realize he was serious, so I shouted back, 'Yes! I'll pull in one of my pigs and my lines, and get a rope and drop a Scotchman overboard.' The pilot circled behind me and I let out a long rope with a Scotchman on it. He shut off his engine and grabbed it and I towed him like a fish. His pontoons were just eight or ten feet from my stern. I rushed inside the cabin and got my scale and weighed out six or seven small spring salmon for him. Then, by stretching as far as I could reach over the stern, and he reaching as far from where he was standing on the pontoon as he could, we transferred the fish and he handed me the cash. The whole transaction took about ten minutes. I asked him where he had come from, and he said he was on his way over from the tip of Vancouver Island, and that he had two other men, a mechanic and a log scaler, with him. They had been out in a logging camp somewhere in the Nootka area on the Island, and while they were coming back they started talking about fish. He said to his passengers, 'It's a quiet day, so maybe I can get you a fish or two from a troller.' 'Do you make a habit of doing this?' I asked the pilot. He said, 'Oh, once in a while, if it's very quiet and I need fish, I come down and buy some.' I untied him then and let him drift back well behind the boat. Then he started his motor and headed off in an easterly direction down the Gulf of Georgia toward Vancouver. And that's the last I ever saw of him."

It began to drizzle. John glanced up at the top of his poles to see if the lines were hanging back or quivering with fish, then looked at his watch. "Nine o'clock," he said. "How about breakfast? On the way in, turn on the big radio. I want to hear the marine weather report and the news."

I set the table, cut a grapefruit in half and sectioned it, fried bacon and put it on our plates, which I was keeping warm in the miniature oven; I had the water heating for tea, and was about to poach eggs in the salted water boiling in the skillet when I thought I'd better take a look to see what John was doing. I stuck my head out the door and called, "Breakfast will be ready in five minutes," before I noticed he had his hand on the gurdy handle and was bringing in a line. He leaned far out over the side with the gaff, hook up, and in a motion so swift that it was only later that I was able to piece together with his help what he had done, he hit the largest salmon I had ever seen, and in a second gesture so fast that it seemed to be all one action he turned the gaff over and drove the hook, whose heavy prong was broadly angled, into the side of the head of the fish. He pulled the salmon—it was a spring—into the boat, grimacing with the effort, and dropped it, with the hook and leader protruding from its mouth, into the fish box. He took his pliers and separated the hook, which was still in the mouth of the fish, from the leader, put a new hook on the line, checked the rest of the leader by running his gloved hand down it for damage, and tossed it back in the ocean to catch fish again. It was all over in less than thirty seconds.

Now, with a quick turn of the wrist, he disengaged the gaff hook from the big spring, and whacked it on the head for a second time, with the wooden end of the gaff just above that hook. The heavy sound of the fish tail thumping against the timbers of the fish box continued for several minutes, slower and fainter, then stopped. Meanwhile, with a deft twist of his pliers, he removed the hook from the mouth of the fish and turned back to the hydraulic control to bring in another leader on the line.

The thud of gaff blows that stunned and killed the fish and the thump of their bodies landing in the fish box became a familiar association of sounds by which I almost unconsciously assessed the quality of the day's fishing, without going out on deck to inquire or count. Thud, thump; thud, thump. On a good fishing day, it was like the beat of a heart in the cockpit, for the steady monotony of its rhythm.

I ran back to the oven, covered the bacon with silver foil, and turned down the stove after peering with a flashlight, the way John always did, down the spout of the kettle to see if the water was boiling. I moved it to a cooler spot on the stove, pulled my red windbreaker over my Stanfield's, and went out to watch him bring in the rest of the lines. John stopped only long enough, when I stood on the other side of the fish box, to smile at me and say, "It never fails. The surest way to bring fish on this boat is to get breakfast ready."

John bent down and turned the wheel, and then was back peering over the side, poised for the next fish. None of the subsequent catch were as big as the first one, but I counted twelve good-sized salmon in the box by the time he had brought in all the starboard lines. Each time he delivered the killing blow, I automatically turned my head and looked the other way. I never did get used to what I always thought of as Death in the Cockpit.

Standing there, gripping the chain rail John had rigged up for me to keep my balance, with my body braced against the buffeting wind, I watched John fish for perhaps a quarter of an hour. The howling wind pushed heavy waves across the windward side of the hull; waves spilled through the scuppers, openings in the sides that allowed the sea to sweep across the deck and out again. John, spray dripping from his glasses, steered and fished in a dual action, busy and serene. Finally, I went inside, but I was barely over the doorsill when I heard a shout and came out again. John was working the port gurdies and was motioning to me. I lurched down the deck as he held aloft by the gaff hook a huge, flat, diamond-shaped fish, mottled greenish-gray on the back, with a glistening white underbody tinged with yellow, and a squinty slit eye. Actually, it had two squinty slit eyes, both on the same side of its head! "Halibut!" he cried. "We'll have halibut for supper. My favorite fish."

I had to stay and watch, and by the time he had gone over his lines he had twenty salmon and the one halibut piled high in the fish box. The giant salmon and the halibut overflowed the checkers they had to themselves; the halibut, even after a second whack between the eyes that stopped it from flinging itself out of the box, continued to make soft, swimming motions.

With all the lines out again, John turned on the hose connected to the hydraulic system beside the starboard gurdies, and washed down the fish with seawater, covered them with burlap, grabbed the knotted rope hanging from the end of the boom, and pulled himself up out of the cockpit. He drew off his rubber gloves, and turned the fingers inside out as he walked down the deck. "I always clean my fish right away, but this time I'm going to have breakfast first and get some dry clothes on," he said.

While I was poaching the eggs and boiling the tea water, he fastened his gloves with clothespins to the arm of the can opener jutting out above the stove. He pulled off his damp Stanfield's and hung it by the sleeves from a line that ran from the top corner of the cupboard to a hook beyond the stove in the ceiling, washed his hands in warm soapy water in the basin, and sat down with a sigh of relief. "That's a good day's fishing already," he said. "My guess is that that big salmon weighs about twenty-two pounds." Later, he lifted it from the fish box and hung it by the mouth on his large brass hand scale. He had guessed right: it was twenty-two pounds one ounce. The halibut weighed seventeen pounds. "Small for a halibut but just right for us," he said. "We'll keep that for ourselves."

As he sat down to breakfast, John reached up to the ceiling and turned on the small citizens-band radio attached there, commonly referred to on fishing boats as the Mickey Mouse because of its short range and limited usefulness. "I want to hear what other fishermen around here are saying," he explained.

I wasn't aware there *were* any other fishermen around to say anything, since there wasn't a single other boat in sight, but a voice came in, loud and clear: "Hello, John, are you there? How is the little lady this morning?"

John snapped off the radio and sat down at the table, scowling.

"Was that your friend John Chambers talking to you?" I asked. "It sounded like him."

Silence, except for the sharp noise of John nervously cracking his jaw. He blew his nose and said, "I've got a very good reputation."

I must have looked astonished, because he leaned over the table, put his hand over mine, gave me a quick, nervous kiss, and said, "I've never had a woman deckhand before, so I have to get used to it, and it will be a surprise to some who know me. You may think we are alone, but please remember that when you are talking to me or anyone on the B.C. coastal radiophone, *all* of B.C. and half of southern Alaska listens with bated breath to every word. We hear some very intimate calls in summer by people on fishing boats or in logging camps and so on—people who are new to radiophones or ex-cityites, who don't realize how many listeners are eavesdropping their love talks, and, too often, their tragedies. Many times, I've heard an obviously alcoholic wife in town and an obviously concerned man, her husband probably, in a logging camp or on a fish boat, worrying his head off as she tries to get dollars off him, half drunk, while he is examining her over the phone about her male drinking companions. You can hear it all in his voice; and then we hear the opposite, too—a wife worried, phoning a boat or logging camp, and the man is obviously high and she's telling him about the kids, the rent that's due, and few or no dollars in the house. So watch out what you say to me on the radiophone when you call me!"

"How do I call you?" I asked.

He said, "If you want me in a hurry, you wire MOREKELP CY3399—those are my call letters; COMOX RADIO, which is nearer home, if it's April, May, or September; or BULL HARBOUR RADIO, up north here, June through August. I can wire back, and within forty-eight hours, if it's what one of my old pals calls an Eeeeeeemergency, I shall go to the nearest telephone."

"How long does it really take?" I asked.

"Oh, it might take one or two days," he said. He spread marmalade on a piece of toast and poured more tea. "How do you like your first fishing day, so far?"

I hesitated, and then, because I had been thinking about it a lot, I blurted out, "Does it bother you to kill those beautiful fish?"

"You need protein, don't you?" he snapped. "It bothers me, but where I hit the fish on the top of the head right over the brain it's killed instantly. That's a lot better than dying in a gill net or seine from strangulation or drowning. I find I never like the killing part. You never get over that. But if I have to kill something, I'd rather kill a fish than a cow, or especially a deer. It bothers me more with a female salmon when they make all that roe for reproduction—especially when the government doesn't keep its promises about salmonid-enhancement programs to protect and expand the stock."

John put on his mackinaw and a dry pair of gloves, which he kept stacked up in one of the three big storage drawers below his bunk, and went outside again. As soon as I had washed the breakfast dishes, swept the toast crumbs from the stove, and thrown them overboard, I followed. The wind had dropped, it had stopped drizzling, and the day was crisp and clear, sunny now but with a sharp, biting chill in the air. It felt wonderful to be alive, and I sat down on my foam cushion on the deck to watch him clean fish. His gloved hands were moving with such a fast, steady rhythm that all the smaller fish, which he told me averaged from six to nine pounds, had been cleaned and he was at work on two large springs. He sent me to get the hand scale again from its hook on the corner of the shelf over the sink, and weighed each of these silvery springs, hanging them by their open jaws. They were fourteen and sixteen pounds each.

"Can you tell the different species of salmon apart?" he asked me.

"I didn't even know there were various kinds," I replied. "I thought that salmon were—well, salmon."

"There are five Pacific salmon species. You'll be seeing them all in B.C.," he said. "Springs are the biggest. We fishermen refer to any springs we catch that are over twelve pounds as 'mild-cures'; they bring the biggest price, because large springs are the finest salmon that can be mild-cured in a soft brine for delicatessen salmon, sometimes called lox. Those two other big springs we caught this morning were average size for mild-cure, but they can run to forty, fifty, even sixty pounds. The biggest I ever caught was fifty-six pounds, but I heard in Namu yesterday about a sportfisherman who came in on a charter boat to refuel and he had a spring that measured sixty-five inches long, twenty-four deep, and fourteen wide. It must have weighed around a hundred and fifteen pounds. The man who sells oil at the fuel dock said he felt sick when he saw it; it was too old to be caught and should have been allowed to die in peace. We've also caught coho salmon today, and ours are all under ten pounds, but they often weigh fifteen, even twenty pounds. Sockeye, the salmon most used for canning until recently, average six to ten pounds; the chums, or what we often call dog salmon, because of their fanglike teeth and hooked upper jaw when they are going upriver to spawn, weigh anywhere from six to twenty-odd pounds. Pinks, which are smaller, average four to eight pounds, have small scales on their silver bodies, and a shorter life cycle. Pinks spawn every two years, compared to four to six years for the other species. Fishermen usually refer to pinks as humps or humpies because they develop a distinct hump on their backs toward spawning time. I'm a spring fisherman, primarily, although I catch a lot of coho, and if I hear there are sockeye around, sometimes I do change the gear for a short time. We don't see any dog salmon trolling, and I try to avoid humpies as much as I can. Too much hard work; one big spring is the equivalent of about forty pinks. I really hate to clean humps. God, the blood! The fishermen call them 'slimers' because they are so slimy to clean. I usually don't take them until a day or two before I go in to deliver, so I can sell them in the round and don't *have* to clean them."

Lifting one of the big springs from the fish box into the cleaning trough and opening its jaws, John said, "Come and look at this." The pigment of its gums was entirely black, even around its sharp teeth. "You can always tell a spring by its black gums, and usually by its spotted back and tail."

I put my hand on the tail; the spots were black, round, and quite distinct.

"The coho has a black mouth, too," he said. He picked up a coho and opened its mouth. "But around the gums, where the teeth come out, as you can see, it's white, and the dots on its tail are not as distinct. Sometimes a spring doesn't have dots but has silver streaks that go right to the end of its tail, whereas the streaks on the coho are more defined and enclosed. The spring has a longer tail to hold it by, and the coho's is stumpier. The back of a young coho is often a deep blue. If you hear fishermen talking about bluebacks, they mean coho. If they are talking about 'smilies,' they mean big springs, because fishermen smile a lot when they catch them. They are worth so much money."

He dropped the spring in the farthest checker, threw the coho back on its pile, and, digging down through the fish, drew forth a longer, streamlined, silvery-blue salmon. "I found this surprise on my gear this morning. It's a sockeye. See how straight the line is along the side," he said. John turned the sockeye over, and in the cavity that he had cleaned out the flesh was a deep red. "Sockeye are plankton eaters and live in fresh waters for a year or so before migrating to sea," he said. "Sockeye's delicious, but I prefer spring. I hardly ever get a sockeye like this when I'm using Number Six hooks; I usually catch them on the smaller Number Fives."

He put the sockeye back and, leaning over, hauled the biggest spring into the cleaning trough again. After he had neatly cut out the gills, using the tip of his knife in a scooping motion, he turned the salmon on its back and made a long slit from its anal opening to its head, reaching in with his gloved hand to draw out the entrails. I had started to turn away, but he motioned me to come closer again and made a second slit in a long sac he was still holding. "I want you to see what this salmon has been eating," he said. "When you see what's in his stomach, maybe you'll feel different about killing. See these tiny things? What do they look like to you?"

"Shrimp," I said.

"In nature, everything is eating something else. The salmon eat the shrimp and other smaller fish, especially herring, and we eat the fish in our turn. It's all part of the natural protein cycle."

A whole flock of gulls were following in our wake, feeding on the salmon guts John had been throwing steadily over his shoulder. After he had carefully scrubbed all the blood from the backbone of the biggest spring and hosed it down—leaving the head intact, as he had done with all the others—thrown it back in the fish box, and covered it with burlap, he turned back to his lines. He was sharpening each hook with a long file that he kept in his back pocket, before he snapped the leaders into the mainline at the three-fathom markers. "Will you steer for a while?" he asked. "I want to go over my gear. Keep on the same course, at forty fathoms, and watch for kelp patches and logs. *Keep looking ahead!* Even when you are out here talking to me, form the habit of *always* looking ahead!"

I went inside. I steered for half an hour or so, looking down every now and then to examine our bird book, because I was trying to identify precisely the little ducks I was always seeing around the boat. I was turning the page with my free hand when I heard a shout from John in the stern, and the wheel gave a sharp twist. I let go of it and watched, fascinated, while it continued to move firmly in the opposite direction to the one I had been steering in. I knew I had done something wrong, so I took a deep breath and went off to the stern to find out what. I waited on my side of the fish box, holding on to the chain, until John leaned back and relaxed. He looked at me. "You went over the pigs," he said. "What were you doing? I'll bet you were reading."

"Well, sort of," I said lamely. "What did I do?"

"All of a sudden, I saw two white floats going past," John said. "I was kind of hazy and thought they were somebody else's, but there's nobody around, and then I looked back and didn't see any floats behind me and thought, My God, those are *my* pigs! I turned the boat as fast as I could the other way and managed to straighten everything out, but do you know what could have happened if I hadn't?"

"No," I said weakly. "You'd better tell me."

"If the trolling wire the pigs are on had gone between the boat and the propeller, it would have fouled the propeller and stopped it from turning and could have cut off the shaft. We would have had to be towed."

I sat down abruptly on the deck. I couldn't think of anything to say except that I was sorry, and this was obvious.

After a while, John said, "Would you like to go in and steer again for me?"

I looked up and saw he was smiling. "Yes, I would," I replied. "I'm going to recommend you for sainthood."

I went back into the pilothouse and steered for quite a long time, until there was a tug on the wheel. I saw the Unseen Hand moving the throttle back and we slowed down. I heard an unfamiliar noise and put my head out the port window. Another troller, white with neat brown trim, was moving slowly alongside us, just far enough away to keep both vessels clear of each other's outstretched poles. The man in the stern was shouting and making motions with his arms at John. Looking back at John, I saw him enthusiastically doing the same thing. The other man held his arms out wide, and held up six fingers; then he narrowed the space between his arms and held up four fingers. In reply, John held his arms out very wide and put up one finger, narrowed them a little and held up two; when he brought his arms in still further, he held up all ten fingers and then another seven for a count of seventeen.

Through the binoculars, I pieced out the name of the other boat: the *Janice T.*; and had a good look at the other fisherman. He was John Chambers. Both boats were rolling back and forth in the waves, while their occupants shouted across at one another as if they were standing on opposite street corners. I could hear John's hoarse voice: ". . . small halibut . . . about seventeen pounds . . . have it for supper . . ."; and John Chambers's voice, coming and going with the wind, say: ". . . six springs and four coho . . . looks pretty dead now . . . going to Danger . . . back again . . . if . . . don't like . . ."

Then John, leaning way out over the stern: "Did you get . . . radio fixed . . . buy new one?"

The voice of Chambers blew faintly toward us on a friendly gust of wind: ". . . Same old one . . . couldn't wait . . ."

Chambers asked a question I couldn't hear, and the words "Goose" and "Goslings" from John's reply hung briefly in the air. The two boats drifted apart; a few minutes later, the *Janice T.* became a mere fleck on the ocean, heading south, and we were alone once more.

"He's going to Pearl Rocks and the Danger Shoals," John said. "If he doesn't find fish there, he'll come back and join us. We met when we were fishing, in 1937. Ever since, we've been looking for one another and exchanging information when we're in the same area. If we talked fish like that over the radio, we might have the entire Pacific fishing fleet with us in an hour."

We trolled along slowly. The day was clear and sunny, the water silky smooth. You could see on its gray surface for miles. Soft white clouds drifted across the sky. "Now that you've got your sea legs, it's time to oil the gurdy blocks; they've been squeaking," John said, handing me a small, grimy oil can with a long spout and pointing to three heavy brass spheres, pulleys hanging about four feet above the gurdies from a davit, a piece of pipe that was bent at the top. The mainlines were threaded through these pulleys out to sea, with a duplicate set on the port side. "Watch for the little hole in the hub of the pulley when you turn the wheel and squirt the oil there," he instructed.

John smiled encouragingly and I cautiously climbed up, shakily holding on to the davit with my left hand while I carefully planted one foot on the gunwale, the other on the board to which the gurdies were bolted across the raised end of the fish box. I wrapped my left arm around the davit and transferred the oil can to my left hand while I turned the pulley wheel with my right until I could see the little hole, about an eighth of an inch in circumference, embedded in the spoke. A roll of the sea and the first spurt of oil from the can dribbled over the greeny brass of the pulley. John reached under the fish box somewhere around his feet and produced a roll of paper towel, tore off a piece, and handed it to me. By this time, the spoke hole had disappeared, so I rolled the wheel again until it came around, shifted the oil can back to my right hand, and pumped in the oil. Mopping up the surplus, I turned, triumphant. "See, you *can* do it!" John said. "Think positive!"

I oiled the other two starboard gurdy pulleys with ease, scrambled down, climbed up on the port side, and finished off the three others. I no longer felt threatened by wide-open space, standing above the deck. I enjoyed the graceful swing of the boat as I looked around from my high perch, holding firmly to the davit, feeling the soft wind on my hair. Even as I stood there, a cooling breeze brushed away some of the sun's warmth, before it picked up speed and whistled through the gurdy blocks.

We were coming along to what I could barely make out was a cluster of diminutive islands, and in the far, far distance the spidery outstretched poles and configurations of other trollers. They appeared to be standing still on the horizon. I counted five vessels. I stepped down and around the heavy black iron hand pump in the center of the stern deck—insurance against the possible failure of the automatic bilge pump that ran on the electrical system. From the time I found out what that clumsy obstacle under the boom was, I periodically fancied myself with great heroism pulling that heavy long handle up and down to keep us from sinking; questioning in a dark corner of my conscious mind: Would I be strong enough to man that emergency pump? I once tried, unsuccessfully, to raise the heavy main fishing poles from their fishing position to their vertical running stance, which John managed with such ease. I could pull up the bow poles, but I kept wondering: What would happen if I *had* to bring the *MoreKelp* in to a dock and could not raise one of the larger poles as well?

John seemed to have forgotten that I should be facing forward to look for kelp and logs, so I sat down to watch him and asked, "When did you start fishing?"

He was putting gear out again and didn't answer until he had finished the starboard side. The sun was high in the sky, and when the wind dropped he took off his mackinaw. He was wearing two short-sleeved T-shirts—a bright blue one over a black one whose edges showed at his sunburned neck and forearms. He was bareheaded, but his eyes were protected by a green felt visor on a dirty white band that circled his head.

He leaned back in the cockpit and said, "I was born in Vancouver, but I grew up on Vancouver Island at a place called Cherry Point, on Cowichan Bay. I had my own little canoe when I was a boy and I did a lot of fishing even then for springs—especially toward fall. One cloudy day, when I was about seven years old, I went fishing with my mother. We used a handline in those days, with very light gear and maybe a half-pound lead. You just twisted the line through a piece of brass like a corkscrew where you wanted the lead to be and then you had to row hard to take the line down. You made a little trigger, a little piece of cedar branch about two feet long that you stuck in upright between the ribs and gunwale on one side of the boat. The stick snapped when you had a bite. A lot of fishermen now have bells on lines to signal a bite. I can't stand the noise of those bloody bells, so I use the springs you see at the tops of my poles that jiggle when a fish bites. It was hard work that day off Cherry Point, off the house there at Cowichan Bay, and we were in a very light rowboat. By late afternoon, we had been fishing for hours in the rain; then we got this enormous bite. We stopped rowing to play the fish, and off it went, towing us in all directions! A couple of times, it came close to the boat—a beautiful spring, about eighteen pounds. It towed us around for a long time while we played it. It took a final run for about three-quarters of an hour, and just as we thought we really had it, the line broke." He paused, and, as if he had passed me a photograph, mind to mind, I saw that little boy in the rowboat with his mother.

"What did you do then?" I finally asked.

"We both sat down in the bottom of the boat and cried," he said. "So much tension had built up, so much excitement. We had not only lost the fish but a very good spoon with it that had previously caught a lot of fish." He stood up and started working on the port side.

"And then what?" I asked.

John shook his head. "I was only seven," he said. "All I remember is how hard we cried, but I'm just as excited by a fish now as I was with my mum in that rowboat or in the dugout canoe I had as a boy. My heart pumps with the thrill of it."

When all the lines were set, John removed the burlap covering the fish he had caught earlier that morning and gave them another thorough wash with the seawater from the hose. "I'm cooling off the fish as well as washing them," he said. "Then I'll put them down in the hold to cool and stiffen before I ice them down. If they are warmer than the ice, they tend to melt it. Ours is flaked ice, and the melted granules turn into sharp icicles that can bruise the fish. Mine are top-quality fish for the fresh and freezer market, so I can't be too careful. They mustn't have any marks." He leaned across the fish box to raise a small board along the rim, revealing an open wooden chute. He slid the fish down it one by one, except for the three largest springs and the halibut, which were too big.

"Where does that chute go?" I asked, leaning over to peer down it. I could see at the other end a pile of salmon sliding every which way with the boat's motion.

"Into the fishhold," he replied. "You want to leave their scales on to protect the fish, so you have to send them down gently. They find their own place slipping around on the ice. The chute saves me taking them down by hand—all but the largest fish."

He handed me the two large springs to carry, which I did by inserting a finger or two in each gill flap beyond the line of their sharp teeth, as I had seen him do. He picked up the largest spring by its gills and the halibut by its tail and stepped over the fish box onto the deck with them. He removed the fishhold's hatch cover, dropped the fish we were carrying one by one on top of the others, and went into the pilothouse. He poured himself the last of the coffee from the pot on the stove, and drank it sitting sideways on the steering seat, resting his legs on the stool, which he had let down. After a few minutes, he sighed, stood up, and stretched, directed me to steer, and went off to ice the morning's catch.

I steered, taking a quick look over my shoulder every now and then to see if the hatch cover was still lying on the deck, indicating he was below in the hold. He was down there so long that I was beginning to worry when, glancing around, I saw his gray head, glasses, and white beard rising through the raised square of the hatch opening. He stood up on deck and glanced at the springs on top of the poles. "Nothing yet," he said, coming in and taking the wheel with one hand, while he scanned the sky with binoculars in the other. "We'll chase birds for a while. There are no fish where there are no birds; where the birds are is where the feed is—usually herring or shrimp—and *that's* where the big fish are."

We followed a north-northwest course away from shore into open water, and soon saw flocks of birds overhead. As we came closer, the sky was so filled with birds that their bodies made a dark border between us and the horizon. Hundreds of gulls, their wings outspread, were skimming above the water and swooping down into the waves—a mass of fluttering wings. Through the binoculars I saw black-and-white seabirds, too, swimming and diving below the surface, their tail feathers like arrows pointed skyward.

John reached for his visor, which he had hung up on the can opener when he came in. "You're seeing herring boiling up to the surface and being chased by the salmon and probably cod. We used to come across schools of herring frequently, but it's rare now. The fishermen have caught them all, the bloody bastards!"

He hurried off to the stern, and I went outside and crept forward toward the bow, gripping the handrail along the edge of the waterproof, canvas-covered roof of the pilothouse outside wall, as I carefully moved one foot after the other sideways along the narrow scrap of deck inside the foot-high bulwark between me and the sea, as I had seen John do when he went to the bow to anchor. Safely arrived at the forward deck, I sat down in front of the skylight and leaned over the gunwale to observe the commotion in the water; its boiling surface every now and then catapulted a gleaming small fish into the air which fell back under again. Noisy screeching and calling from overhead birds, who scattered as the *MoreKelp* passed through the splashing turmoil.

I thought I heard John calling. I painstakingly reversed myself along by the pilothouse windows, stepping gingerly over a heavy chain that blocked easy return to the stern deck. I was conscious of the steady hum of the gurdies doing their job, and the thud-thump of successful fishing. John waved impatiently, so I hastened back to see what he wanted. He held up a shiny fish about six inches long, alive and wiggling, above the open slit in the abdomen of a large spring salmon. "Here's our lunch!" he exclaimed. "Boy, will that be delicious!"

"What is it?" I asked.

"It's a nice fat herring," he replied. "We're in luck. I pulled it from this big fellow's stomach before his digestive juices even had a chance to start working on it! Well, maybe just a little."

I was horrified. I looked at the wiggling fish, then at John, to see if he could possibly be serious. He was laughing. I said, "You're kidding, of course."

"You'll change your mind when you taste it," John said firmly.

He cleaned the large salmon and the herring, then the other fish he had caught which were still to be done, washed them all down, set the herring aside, and covered the others with burlap. He ran the hose over the cleaning trough, which was covered with blood. Then he picked up the herring and went into the pilothouse. He turned up the stove, got out the frying pan, and handed me two slices of bread to toast over the burner. He scraped the scales off the herring, and while he was rolling the fish in cornmeal he looked at me and laughed again. "With all your cookbooks, I'll bet you don't have a recipe for predigested herring," he said. "They probably never heard of it."

He sautéed the little fish in safflower oil, split it, and plucked out the backbone, which separated easily from the cooked herring; put half a fish on each slice of buttered toast on our plates; and sat down. He ate a piece of herring and held out a second one on the fork to me. I shook my head and backed away. He ate another segment, closing his eyes with a blissful expression. He opened them and said, "What's the difference between eating this and a salmon or snapper?"

"It's been in someone else's stomach first, that's what!" I replied. "It's the idea. It makes me feel funny."

"Don't be silly," he said, squeezing a piece of lemon over the remaining herring. "You eat raw live oysters, and this is very nicely cooked, if I do say so. You'd better hurry up if you want any."

I shut my eyes, opened my mouth, and he popped in the herring and buttered toast. It was delicious. "Marvelous," I said. "I don't suppose you'll ever find another one like that."

"Oh, once in a while if we're lucky, we'll accidentally catch a herring on a hook," he said. "This one's just been swallowed; it still had scales. You can't eat the herring if the digestive juices of the fish that ate *it* have really started working."

John's method of herring fishing was an eccentric dimension to our trolling life. He caught another several days later, this time in the mouth of a coho he pulled out of the water. Only the tail of the herring was visible, and when John retrieved it from the coho's throat it was in perfect condition.

We finished our lunch that first day with cheese, Britl-Tak, and a bottle of beer that John kept in a cache in the space at his feet under the fish box instead of on ice, because he didn't like cold beer. He continued to fish until two o'clock in the afternoon, when he pulled in all his lines and stored the gear away. Promptly at two-thirty, he turned on the CBC radio to the concert hour. He switched off the motor and we drifted while he lay flat on his back on the deck, a sock over his eyes, and, until he fell asleep, listened to a piano concerto for left hand by Ravel, with Leonard Bernstein conducting.

It surprised me that we could pull in the gear and float like this safely, with no anchor, but we were well away from the rocks of the coast or the shipping lanes. I always thought of that two-thirty to three-thirty semiconscious peaceful respite as the Sacred Hour. It was a necessary pause for John, to take a quick sleep after five or six intensive hours of fishing that began between four-thirty and five in the morning and continued until after sunset, averaging from twelve to fourteen hours a day, rain or shine, in the cockpit.

Unless we were running or it was calm, the stabilizers, two flat heavy galvanized-iron plates looking a little like torpedos, were hanging from ropes halfway out along the main poles on either side, their three-foot surfaces horizontal to the sea, two to two and a half fathoms below the water. When we drifted like this, with the gear up, the groundswell would turn the *MoreKelp*, which was heavier in the stern than the bow, at a favorable angle into the waves—from forty to sixty degrees—rather than rolling it into a trough. The sea motion set up a tranquil swaying that was comfortable and soporific.

That day, I was too curious about everything to sleep, and I lay on top of the hatch, holding on to a rope dangling from the boom so I wouldn't fall off, and admired our boat's superstructure piercing my view of the sky: the intricate webbing of ropes, poles, chains, and lines; the tall mast, with its fragile ladder running up the rigging to the crosstree; the small square of shiny metal that was a radar reflector at the tip of the port main trolling pole. Unlike almost all other fishing boats, the *MoreKelp* was not equipped with radar. Wooden boats don't necessarily show up on radar, so without this aluminum reflector we might have been invisible in a fog. Our reflector appeared as a blip on the radar screens above the steering wheels in the pilothouses of nearby fishing boats, tugs, cruise ships, freighters, log carriers, and other large vessels where radar is standard equipment. It had a second useful purpose: I could identify the *MoreKelp* among the mass of boats tied up at docks in busy harbors like Port Hardy or Namu when I had gone ashore to shop, do laundry, telephone, or visit friends. Instead of having to pace up and down ramps searching, I could find it even if John had moved, as he frequently did, because he always made a point of tying up alongside the outside boat for a quick getaway. I would look up and across the tops of all the trolling poles until I sighted our distinctive shiny metal square, set up on end at the tip of our port main pole. Our poles were also different; they were painted green among a bevy of aluminum, natural, varnished, or white ones.

Precisely at three-thirty, when the music ended, John arose from his nap, donned gloves and eyeshade, and went back to work. Toward the evening, I had my first taste of his shoal-and-reef fishing—what he and other fishermen called "rock-pile fishing"—along the underwater ledges where spring salmon congregate. We arrived at the rocks of Goose Island and its satellite islets, the Goslings, at slack tide, late in the afternoon. Slack tides are favored for fishing—especially if morning and evening slacks coincide with changing light. We picked up fourteen coho, four sockeye, and three big springs while the sun was setting.

No steering was entrusted to me among those dangerous reefs, in that wild surf that rolled the *MoreKelp*, so that one minute I was gazing at a tipped view of snowcapped mountains, patches of blue sky, and tree-bordered shoreline, and the next looking deep into the dark waves and the shadowy rock edges underneath. We trolled back and forth in a half circle, with the sounder plunging to sixty fathoms and leaping up to twenty, dropping to thirty and then—hold your breath—rising to ten for a single flash before the Gardner roared as John revved it up and swiftly moved away from jagged underwater peaks. Sitting forward in the pilothouse, instructed to keep looking straight ahead for logs and seaweed, out of the corner of my eye I saw the orange light of the pilothouse sounder flashing 40, 30, 20, 18—would it never stop? I pulled out a chart to locate our position. We were fishing a rounded edge, it was plainly marked on the charts: full of reefs, and the fathom markings read 64, 18, 31, and 10, in that order.

Clutching the edges of the counter to steady myself, I walked outside and stood by the gurdies, holding tight to the chain rail. I alternately looked ahead and watched John maneuver in and out among the rocks and pull in fish, in a sunset that threw a glow across the mountains, across the water, across John's face, setting off a fiery gleam from his eyeglasses. He finished with the mainline, and before he started on the pig line he stopped to remove his visor and wipe the spray off his spectacles. He gave me a radiant smile as he put them back on again and grabbed the wheel to turn into the pounding waves.

I leaned over and shouted, "Don't you ever get scared?"

"I love it!" he shouted back. "I've been steering this edge for thirty-five years and I *love* every minute of it!"

The wind died as he pulled in the last spring from the last hook on the second pig line, and while he was stowing away the gear for the night in the forty-five-minute run into the Goose Island anchorage, I went into the bow and surveyed the canned-goods supply, to see what was available. Besides a liberal supply of evaporated milk, there were mushrooms, and there was every kind of vegetable—corn, peas, carrots, tomatoes, beans, beets; and a wide variety of fruit—apricots, peaches, raspberries, strawberries, applesauce, gooseberries, and cans with exotic labels that read "Tropical Mixed Fruits." I looked around and found several tins of butter, which I remembered from our winter trip as being even better than the fresh; sardines, clam chowder, tomato soup, consommé, corned beef, pork and beans, bacon, kippered herring, tea, coffee, cocoa, and tucked away in a corner, two cans of Spam, which I pushed as far into the bow as my arm could reach. There were enough canned goods—John called them "unfood"—for several months. John had said he wanted a bowl of fruit topped with yogurt and wheat germ, so I selected a can of apricots and climbed back up into the pilothouse. John was standing at the wheel. "Steer while I ice the fish I just caught," he said, his voice strained and tired. "Just keep the bow pointed toward that little bay, and I'll be back in time to take us in."

I took the wheel and did as he directed, but what was taking him so long? We were coming in so close to the beach I could see the trees separately, the silvery pale driftwood and the brown kelp, rolling and broiling among the rocks along the shore. I felt the edge of panic, and timidly moved the throttle back and slowed us down. John reappeared immediately, pushed it to full speed again, and we came in much farther toward shore than I would have dared. It was months before I could judge distances over-water, and I was never really sure. When I steered, I allowed miles between us and other solid objects—especially moving ones—and was still uneasy. In poor visibility, I sometimes couldn't tell whether a vessel was coming toward us or going away until I had a good look through the binoculars.

We had the little harbor all to ourselves. I have always been around people, so I assumed that I would welcome the comfort of other boats, but, like John, I found I was happiest when we were alone—as if this whole exquisite landscape had been arranged for our pleasure. John had favorite protected harbors in the lee of islands or the mainland, and we anchored every night or two in a different place. Each sheltering inlet had its distinctive coloring after the sun had set. Possibly because it was the first of which I was conscious, the Goose Island anchorage in the lee of that low rocky promontory was my favorite, and we returned to it many times over the years.

Light faded mysteriously, vividly, in these protecting harbors. In Shadwell Pass, in the lee of Hope Island, on the edge of Queen Charlotte Sound, one night at nine o'clock in the evening, I wrote in my journal: "The sky is liquid blue, with a great orange stripe across the horizon that makes the world orange"; and another night on that same trip, at another favored spot, Grief Bay, at the south end of the much larger Calvert Island: "This is a dark-blue-black-and-silver harbor." But the soft lavender and silver tones through the trees at the Goose Island anchorage were for me the loveliest of all.

When the ritual of anchoring had been completed, back and forth, back and forth, until the anchor held firmly enough on the bottom to satisfy John, he came inside and sat down on the steering seat, picked up from the compass shelf a pen, and his worn five-by-eight-inch brown leather book with the words PILOT HOUSE LOG FOR FISHING VESSELS stamped on the inside pages, and began to write. (Each evening on the boat during the fishing season, he made an entry, and that moment of quiet writing marked the end of the working day.) As the light faded into night, we ate a light supper, prepared gear for the next day, listened to weather reports over the radio and the long-distance conversations of other fishermen to their families on their VHF (very high frequency) radiotelephones, and slept, in that order. Bedtime was rarely later than 8 p.m. Going to bed with the onset of darkness and getting up before the sun suited us so well that I wondered why I hadn't always lived that way.

In the logbook, which had "With the compliments of INTERNATIONAL PACIFIC HALIBUT COMMISSION" inscribed on the inside cover, John entered the date, location of each night's anchorage, type of fish he had caught, whether he rested, and the number of hours spent in the cockpit, in a personal shorthand that told the story of his day in a single line: Ju 24 Goose Id. 6 Big [14] 15 c. R 1 Hal. 4 Sock. ⑮.

I asked for a translation of his hieroglyphics. "The squared numeral is the number of springs I caught today; "c." is for coho, "R" for rest, "Hal." for halibut; Sock. for sockeye, obviously; and that circled figure at the end is the number of hours I spend each day running and trolling but not working on the gear," he said. "Working on the gear is extra;

I do that in the evening or when the weather is bad and I can't get out fishing."

While John was taking his pre-bedtime nap, sitting up with one shoe off and his head bobbing on his chest, I picked up the logbook and began leafing through it backwards—an unfortunate habit of reading I got into years ago. A month before I came, he had fished at Goose Island for two days with a score of [12] 3 Big 50 c. (12½) for the first day, and [11] 3 Big 103 c. an R and ⑮ for the second. Entries farther back also included "Dog" for dog salmon and "Hu" for the lowly humpbacks, and, very occasionally, "Cod." A noticeably reduced catch was often followed by the notation "Fog," and occasionally the word "Skunk" appeared—the colloquialism for failure to catch any fish at all that day despite spending ⑪ or even as long as ⑭ hours in the cockpit. Mechanical and oil changes, renewal of lines and major repairs were recorded in red ink, and the death of his father at ninety-seven in Montana on the previous April 25—*Heywood Died Billings*—was firmly written and circled with green ink, followed by a hiatus for one week in entries while he made an unprecedented interruption of his fishing season to attend to the funeral and his father's affairs.

Continuing my backwards reading, I found he had marked the beginning of 1971 in red ink as the "20th Season for the *MoreKelp*," and had noted in the final entry for that year major recent changes in his boat: a new stern and pilothouse, a new anchor winch and hydraulics, an automatic bilge pump, and two new water tanks with a total two-hundred-gallon capacity—an unusual amount for any boat of that size, for which I was increasingly grateful. Unlike other fishermen, who had to conserve their freshwater supply, he was always urging me to "use more," so he could freshen up at the next watering hole—usually Namu or the oil dock at Port Hardy.

Back in the nineteen-sixties, I found occasional quotations that he must have wanted to remember inserted among the hieroglyphics: "Give your husband a few tools & in no time a dripping faucet will become a running stream" was penciled in at the top of a September page in 1966, and he used up half a page in June that same year with the Shakespearean quotation, in red ink:

> This most excellent canopy, the air, look you,
> This brave o'erhanging firmament, this
> Majestical roof fretted with golden fire

> —Why it appears no other thing to me
> Than a foul & pestilent congregation of
> vapors.
>
> Hamlet (Re: Air Pollution)

The first entry in this logbook, which began in 1964, was a half-page listing in a different, neater hand, entitled *Sean's poundage landed for June 1–Aug. 18*, with a picture of a small, openmouthed salmon drawn along the margin, which was decorated with dollar signs. John's younger son, Sean, who was twenty-one then, had listed the salmon he caught in eighteen gradations, adding up to 20,475 pounds of salmon; 1,073 of halibut; and 303 of lingcod during that period. I had just come to the final entry in the logbook, a quotation John had copied onto the inside cover, by Honor Tracy, that he had liked so well he had also inscribed it on the pilothouse wall above the stove—"The time was between 3 and 4 in the morning, when the blood flows slowly and the future appears as a likely succession of defeats and disasters"—when John opened his eyes, groaned, and turned up the radio to the eight o'clock news. Five minutes after the news and marine weather report, he was in his bunk, red blanket pulled up to his chin, his mackinaw hung at the window to shut out the light, and his heavy wool pants folded on the steering seat. He was fast asleep while I was still gathering up my belongings to go down below.

VI

The alarm went off at four-thirty the next morning. I waited to get up until I heard the Gardner start and the clatter of the anchor chain going up over the side of the boat, and knew we were on our way to the fishing grounds. It was dark when I appeared in the pilothouse. I brushed my teeth and washed without turning on the light, so that its reflection off the windows wouldn't interfere with John's ability to see debris or logs on the glistening black, moonlit water. He stood up while I slid past him on the bench into the corner, and he handed me a cup of hot coffee. Then I heard him rummaging under the mattress of his bunk. He returned with an elegant but worn, long, thin volume bound in cream-colored vellum, and handed me a flashlight. "Hold this on the book, so I can read my favorite poem to you," he said. "We're going to have a wonderful sunrise."

The first streaks of yellow light blazed across the horizon while the moon still cut a cool white path of phosphorescent froth on the water. He began to read in his gravelly monotone:

> Wake! for the sun, the shepherd of the sky,
> Has penned the stars within their fold on high,
> And, shaking the darkness from his mighty limbs,
> Scatters the daylight from his burning eye.

He continued reading, and when the gray light filtered in through the pilothouse windows, I turned off the flashlight. Several stanzas, several pages later, he closed the book, laid it on the compass shelf, and said, "I'll always keep my balance, the way my dad taught me in regard to drinking, as I don't want it to affect my brain, but I will reserve the right to get drunk in high altitudes and on the ocean occasionally and shout and howl at the sunset and the moon, and shout Omar Khayyam at the stars." He pulled the throttle back to the slower trolling position and went on deck to put out gear and start his day.

I picked up the slender volume. It was, of course, the *Rubaiyat of Omar Khayyam*—not the familiar FitzGerald translation but a paraphrase, by Richard Le Gallienne, published in London in 1897 on the finest rag paper, with "Bache Cottage, Chester," written in black ink on the cover, and inside, in the same hand, the name Muriel S. Daly, John's mother.

When John returned a half hour later, he was carrying a galvanized-iron pail under his arm. "Washday," he said cheerfully. "It looks as if it's going to be a sunny day, so everything will dry outdoors. That'll use up some of those two hundred gallons of water, so when I refill the tanks in Port Hardy it'll *all* be changed and fresh."

"I haven't anything to wash yet. I just got here," I said. Hand-washing heavy clothes in a pail! I didn't even know how to go about it. "Wouldn't it be a lot easier to wait until we get to Port Hardy or back to Namu and find a laundromat?"

"Nonsense!" he replied. "Unlike most fishermen, who want to conserve water, I have so much, I always wash clothes on the boat and encourage others to do so; it makes good elbow grease. One of the reasons I'm so fond of my heavy Norwegian socks, fish pants, and Stanfield's is because they are all large loose-weave wool and wash easily, with a rinse. I love to be alone after a trip to Namu or Port Hardy, but when I'm in port I want to talk and socialize to the maximum, not lose talk time washing clothes. Besides, I took my sheet off my bunk this morning. I change my sheet for a new one every month whether it needs it or not." All the while he was talking, his right arm was vigorously pumping water into the bucket in the sink. He set the pail, three-quarters full of water, on the stove, turned up the burner, and said, "Let's have breakfast as soon as I get the gear out, if we don't get a bite," and was gone.

Fishing was slow. Around noon, when the sun was high in the sky and it was warm enough to sit on deck without a sweater, he took his bundle of laundry, a bag of clothespins, a jar of detergent, the bucket of hot water, and me on deck, rolled up his sleeves above his elbows, and plunged his sheet, our dish towels, his dirty pants, wool shirt, Stanfield's, and socks, one after the other, into the soapy water in the pail. With the same enthusiastic pumping motion he had used at the sink, he scrubbed each item, his elbows churning up and down above the top of the pail. He offered to wash my clothes, too. I looked down into the black water and shook my head. "No . . . no," I said. "I'll do mine when I need to later." He rinsed everything in a lukewarm mixture of water from the kettle and the sink pump. He wrung out the sheet and threw it over the boom, pinning down the two ends below; distributed the dish towels, Stanfield's, pants, socks, blue T-shirt, and white Norwegian mesh underwear in a decorative line along the chain above the starboard gunwale; and fastened them by doubling their ends below the chain as he had done with the sheet, so they wouldn't blow away.

I was dazzled by John's simple efficiency.

"I love that fresh smell of sea air in clean clothes," John said. "You can't get that in a laundromat."

Our laundry made a sharp flapping sound in the wind that was beating them dry: red, white, blue, yellow, brown, and gray semaphores—signals of our industrious housekeeping.

By silent agreement, I gradually took over all the traditional female household chores—especially cooking and laundry. It was what I knew how to do, and I wanted to be useful. I was rescued from the washing-pail routine by the size of our laundry, with two of us on the boat: more towels, sheets, clothes—more everything. My first move whenever we came into port was to take a bath and wash my hair, and my second, to locate a washing machine.

John had put on the automatic steering device while he did his washing. He poured the last of the wash water over the rail, looked up at the sky, and said, "I'm going to bring in the lines and we'll go ashore at Goose Island. It's a gorgeous day, and we may not be so lucky again." We were richer by two springs—one a big mild-cure—and five coho by the time he had the gear away, and after he had given the fish their seawater rinse and covered them with burlap, he came inside and steered close in to the lee side of the largest of the islands around which we had been fishing.

The islets were scarcely more than rocky piles sticking out of the water, while Goose Island, which looked to me to be about a mile long, was a finger of land with a growth of fir trees down its spine, encircled by a sand-and-rock beach jammed with driftwood that the tide, sun, and wind had worn down into bizarre shapes.

John changed his boots for a pair of rubber-soled shoes and went outside to drop the anchor. When he was satisfied that it was holding, he went back to the stern, gave the fish a second wash, and put them down on the ice. We had a quick lunch, and I was cleaning up when I heard the pulley creak that meant he was letting down the dinghy and sliding it into the water. I ran out to help, but he was already standing in the dinghy alongside, holding on to the *MoreKelp*, waiting impatiently for me to climb over the gunwale and get in. I groped awkwardly with my foot on the steep side for a foothold, gave up, and slid instead into the dinghy and sat down. He cast off and started rowing with his quick, choppy stroke toward Goose Island through booming surf.

"This is our first trip ashore," I shouted happily.

"Goose Island is a bird sanctuary and I want you to see it," he shouted back. "Watch out for the gulls; they'll dive-bomb you to defend their nests."

He maneuvered the dinghy among the rocks along the rough bottom, and since I was wearing sneakers but no socks, I jumped out, sloshed through the water, and pulled the dinghy by its painter—the long rope attached to a ring on the bow—up onto the shore. John stepped out on the shore, took the rope, and secured it around a heavy rock at the tree line, above the high-tidewater mark. He said, "Keep your eyes to the ground and don't step on any nests."

We walked up from the shore onto the big rocks. Instantly, gulls rose right out of the ground, from everywhere; we were surrounded by screaming birds, flapping their wings, swirling around our heads. I turned and looked back.

Between us and our vessel, rocking placidly at anchor, were the twisted shapes of the pearly-gray driftwood and, beyond, brown kelp churning in waves pounding the shore. Gulls swarmed over us, their cries increasingly strident. Looking down, I saw a nest at my feet, with tiny beaks of new babies poking out, then another and another. "Let's get out of here," John shouted, and we turned and scrambled over the rocks to get back to the dinghy. On the beach, he reached around a wind-beaten log with a thick, contorted branch that curved straight up from a heavy, gnarled trunk base like a sea monster at rest, and picked up a green clear-glass sphere a little larger than a tennis ball. "This may have floated over from Japan, from a Japanese net float. Our lucky day." He smiled and handed it to me.

Hurrying back in the dinghy, clutching my glass ball, all I could think about—it was a huge fear, eating away inside—was the ordeal I faced: the climb back onto the boat. No problem at all for John, with those long legs, to step from the dinghy seat over the side, but I was still haunted by the memory of my desperate scramble to get aboard when I arrived at Namu. The same graceless scene was about to be repeated—no way out. Paralyzing anticipation wiped out the joy of our excursion.

We touched gently against the planking of the *MoreKelp*, and John said pleasantly, "Ready?"

"I can't," I said miserably. "I just can't get up there. It's too far."

"Come on now," he urged softly, but there was no mistaking the sternness of his tone. "We'll try a different way. Stand up on the seat and hold on to my shoulder, and I'll make a step with my hands for you to stand on."

The dinghy rocked in one direction, the *MoreKelp* in the other. With a mighty effort to surmount my fright, I held on to John's shoulder with one hand, stepped into his locked fingers, and he slowly heaved me up and over the side. I scrambled to my feet as he came aboard, and while we were pulling up the dinghy and fastening it down, I said, "I know what I'll do. I'll get myself a little ladder to put over the side when we go ashore. What do you think of that?"

"Certainly not!" he exclaimed. "I have never seen one on a commercial troller in all my years fishing. What do you think you are on—a yacht?" He patted my head as if he were soothing a small child. "We managed quite well, didn't we? Think positive!"

I made a silent pledge: no more excursions by dinghy off the boat.

While John took his afternoon music nap, I lay in my favorite spot on top of the hatch and considered my problem. The answer: a ladder, no matter what. At three-thirty, John got up, turned on the motor, and began putting the gear out for the afternoon bite while he steered away from Goose Island and the Goslings out to sea. We moved slowly toward the place where I had seen five boats in the distance when we had first approached Goose Island—was it only yesterday? I had already lost all sense of time. I counted seven trollers there now, poles outstretched, fishing in a tight cluster, far removed from any land mass in what appeared to be open ocean.

"Why do they stay so close together when they have the whole ocean to choose from?" I asked.

"They are fishing along the edge on the Goose Island Bank," he said, showing me on the chart a roughly shaped ellipse of shallow reefs between seventeen and forty fathoms below the surface, where the boats were. "It's a spot favored by salmon—particularly springs."

When he had all the lines out, John said, "I'm going to ice this morning's fish now, so I'll put on the automatic pilot—we call it the Iron Mike—and show you how it works. It's very useful when you are traveling in a straight line, which is monotonous. The Mike leaves you with both hands free to write letters or make entries in a logbook, or cook—things like that. You still have to keep looking ahead for logs and kelp and pay attention—especially with my old Woods–Freeman Mike, because it sometimes slips. The biggest danger is falling asleep and running up on the rocks."

We went into the pilothouse and I sat down on the seat. John held the steering wheel with both hands, turning it back and forth, so that the boat moved a bit to the right, then to the left, until he had it in a set rhythm, going forward in a relatively straight line. He turned on a switch near my knee to give the Mike power, then leaned across my feet, which were resting on a raised platform, to reach a piece of tarred line that I had never noticed, whose end disappeared through a hole in the wall behind the steering wheel. It was attached on the other side on the open panel behind my bunk to a clutch, among the oscillating chain belts and gears. The end on our side was looped to a hook on the wall at my feet. He took the string off the hook, gave it a yank, and said, "That engages the clutch, and the Iron Mike is on. When you want to go back to steering by hand, you hook it up again to take the boat *off* automatic steering. What's the matter?"

"It is so illogical," I said. "I would think you'd put the rope *on* the hook to turn the Mike *on*. I'll have to write it down or I'll mix it up."

"You haven't been around motors and mechanical things, living in a New York apartment," he said. "You'll have to take my word for it; or else go down and ice fish for me, and *I'll* watch for logs and kelp instead. You could do it easily. My sons iced fish, and all my deckhands did, too."

He disappeared and I sat watching the wheel but not touching it, fascinated with its turning back and forth in a steady beat that kept our bow pointed on a west-southwest compass course. I added the Iron Mike to my list of unseen boat companions.

I soon began to yawn, lulled by the easy sway of the vessel. Doggedly staring ahead, I could feel my eyelids drooping. I leaned down and put the rope back on the hook and grasped the wheel. I turned the boat—too sharply for the gear trailing behind us. I hastily put us back on course and was still steering by hand when John reappeared and took the wheel. "I wish you'd learn to ice fish," he said.

"I'll think about it," I replied, "but I froze three fingers years ago when I was reporting a story in the Canadian Arctic. They still hurt a lot when they get cold. Anyway, I don't think I'd like to be down in the hold with all those dead fish."

"Speaking of going down in the hold," John said. "There was this fisherman who had his wife on board, and they had a fight, so he put her down the hatch and fastened down the cover. When he came into the fish camp—that's a place run by a cannery that buys fish and where fishermen pick up ice and supplies—the guys on the float heard this pounding, pounding, and one of them said, 'Is somebody down there?' and he replied, 'You're bloody well right there is, and I'm going to damned well keep her there until she apologizes and behaves herself.' "

John put the wheel back on automatic, and was heating water for tea while he spread peanut butter on Britl-Tak. "What happened then?" I asked uneasily.

"Oh, I guess eventually he let her out," John said cheerfully. "Don't you think he had a good idea?"

"No, I don't," I said. "I don't think I want to ice fish down in your hold, either."

We laughed, and John pointed to a group of eight little black-and-white diver ducks bobbing along on a log off to the side of the slowly moving bow of the *MoreKelp*. The ducks shifted weight to keep their balance in the waves of our wake. We waited for them to dive, but they all stayed up on the log. "They're having a discussion," John said.

"Why don't we stop and fish here?" I asked. "I thought birds were a sign of fish."

"I want to try the nearest edge—the Goose Island Bank over there—that's why," John answered. "We've come about fourteen miles from last night's anchorage, and the Bank runs to fifty miles offshore. Fish gather at the edges of rocks under water, especially when they come out of shallows. I've gotten some enormous springs around there in years gone by." He was looking at the other boats through his binoculars to see if the other fishermen had fish on the lines they were pulling in off their sterns. "My old friend Kal Kaisla used to hide behind his pilothouse windows when he was watching through his field glasses like this. He didn't want his friends to see him looking at them. Kal had a lot of children, and he always took his family fishing with him. He had those little nippers of his all over the boat. I asked him once if they helped him and he said, 'Much more entertainment than help. They make toast all day, and then they walk on it with their rubber boots and make a terrible mess.' "

We arrived inside the circle of offshore trollers in time for the afternoon bite, moving slowly, keeping well back from the vessel in front of us to stay clear of his extended pig lines off the stern. On closer view, the circle was a wide one, encompassing several miles of ocean—plenty of room for us all as long as everyone traveled at the same slow speed and kept carefully away from the stretched-out poles of neighbors on the turn. I had started a needlepoint eyeglass case for my mother, with strawberries against a black wool background, and I spent most of the time on the pilothouse seat, shifting my eyes between the velvet-gray surface of water, with its rolling swell, watching for kelp and logs, and my needlepoint. Occasionally, I traveled back to the stern to announce the presence of a clump of seaweed ahead and bring John tea, always glancing up at the springs on top of the poles for signs of a bite, the way John did. Back in the pilothouse, I listened to the ticking of the flasher-sounder, like a metronome, reading, with its staccato voice, our route over the bottom of the sea around the edge of the Goose Island Bank, twenty to forty fathoms down.

On the recorder-sounder he had installed in the stern at eye level over the fish box, John could actually *see* the outline of the bottom while he fished. This sounder was an oblong green box about eighteen inches high and a foot wide, with a glass front through which he could watch a small recording needle travel across a roll of white paper, translating depth soundings by electrical impulse into jagged black lines that looked like those on an electrocardiogram and gave the story of the underwater terrain over which he was maneuvering the boat, its lines, lures, and hooks. The recorder-sounder even showed the presence in linear blips of schools of fish—feed for the larger fish he was seeking. When I went outside again, I asked him to explain it to me. "It tells me what the bottom's like, where the feed is, and the depth," he said. "I try to buy all my electronics from a group of handicapped workers who started their own electronics firm in Vancouver. Mine are not the most up-to-date models, but this one is big and easy to monkey-wrench; you can look back on the paper roll and see where you've been. Before we had electric sounders, we could only guess at the depths or take soundings with a handline or use the fishing gear itself and keep a close record of landmarks in our logbook. We older fishermen are lucky, because we can catch fish without the latest electronic gear, while all the young ones are dependent on expensive equipment. They *have* to have all the dooflaps, like radar and loran, which is a machine that tells you your position on the water through radio-beam signals from fixed positions on the coast—another object to clutter up the pilothouse. I wouldn't know where to put one." He leaned over and adjusted a knob on the box. "I'm not happy with modern machinery," he said. "I have less electrical equipment than eighty percent of my fish competitors, and my net dollars are above average. Those fellows are slaves to consumerism; the upkeep hours for electronics are endless. What's worse, you lose confidence in *self* with too many electrical aids, and you lose *challenge*."

We were making the turn to head in the opposite direction when I saw a familiar-looking troller passing at the other end of the wide circle. "Isn't that John Chambers?" I asked. "When did he come?"

"About an hour ago," John said. "If you had been looking ahead for kelp and logs instead of sewing with your head down, you would have seen him."

Back in the harbor, after a long run that night, the two fishermen exchanged catch sizes and other fishing information again, leaning over the sterns of their vessels to communicate. "When we were younger, one of us would have rowed across to the other," John remarked later. "But after seventeen and a half hours in the stern, fishing, and the long trip in today, I'm just too tired, and so is he."

During the night, I had heard the wind singing in the rigging and felt the waves lapping against the bow, and my bunk rode up and down like a teeter-totter. The next morning, there was a terrific groundswell, and John said, "It's blowing offshore someplace. I couldn't get the weather report from Bull Harbour when I woke up, because a boat called the *Sea Fantasy* is on the rocks up near Prince Rupert and they are holding the channel open."

"Two-foot chop with moderate swell" was the weather report when he did get it, at 8 a.m. John sighed. "A fisherman has to make a hundred and fifty decisions every morning before the sun has properly risen," he said. "A fisherman, or a fisherwoman, as you will become, has to be able to change plans completely five, six, or ten times a day, or an hour, or a minute. There are too many toos to this fishing—too hot, too cold, too wet, tides too high or too low—and flash decisions can make or break you. In summer, the movements of fish rule all decisions, and I will often go one hundred miles from where I expected to be. Never be surprised, my dear, even though you will hate it when you are planning to meet someone at Port Hardy or Namu, to find that I have suddenly taken us off toward the opposite end of the compass. I find that my best decisions are made after a good night's sleep, and my worst usually under great exhaustion."

I gave him his coffee and toast while we were running out to the fishing grounds at the Goslings again. After the weather report, he remarked, "We have an excellent weather service that we should be listening to constantly. It gives three predictions a day, and the local weather four times daily. I don't miss very many. All day long, you are sizing up predictions against what is actually happening in your area, and you choose a place to anchor at night mainly on the predictions. Usually, I try to find an anchorage twenty minutes to an hour away, but I'll run two hours to a good harbor if a heavy wind is predicted. Most harbors are one-wind anchorages; you have to judge what points and islands protect you against north and northwesterly winds, and others that are south or southeasterly. Sometimes you have to dodge around islands, but you want to avoid making mistakes, because you have to move then in the middle of the night. You must never get caught on a lee shore; that is, in a wind that blows you onshore. I always anchor so that the wind blows me *off* the shore, and then if I'm blown anywhere, it will be out to sea, which is a lot safer than being blown on a beach. When in doubt, I start the engine and idle it and make tea, but I don't lie down again properly, because if the wind changes I want to be ready to pull out again quickly. That means up in a flash, and my starter is at one end of the cabin and the hydraulic controls for the anchor winch are at the other end, so I have to move *fast*."

He held out his mug for more coffee. He was wound up in his subject and kept right on talking. "The safest thing to do, of course—and the tough guys who can sleep anywhere do that—is to just anchor wherever you are offshore and roll. There's nothing but ocean and other boats to run into, and everyone's up anyway if it's blowing really hard, so you doze with your clothes on. I can't take it anymore, because

I have to have my sleep now, but I remember once, when I was twenty-five miles out on one of the big banks off the west coast of Vancouver Island, it started blowing hard about 1 a.m., and at two-thirty I started my engine just in time to have to go right over on my anchor line to dodge a fellow who went flying by. Lucky I didn't lose a pole!"

John Chambers came out of the harbor, running right beside us, and we all settled down to the familiar slow trolling regimen. I cooked breakfast, washed dishes to the undulating rhythm of the ponderous groundswell, hanging on to the stove and counters to stay on my feet as our gently rocking *MoreKelp* was transformed into an endlessly rolling tormentor. Staying upright was an exhausting effort, and when I had the dishes safely away and had jammed the knife in the cupboard door to keep it closed, I staggered outside to see how John was managing. I gripped the pilothouse doorframe, the top of the hatch, the mast, the starboard safety chain, on my way—one hand after the other. "Four big springs!" he shouted, leaning over the side with the gaff and hauling in a fifth. I turned my head away when he hit the salmon with a solid whack on the head. I could hear the desperate reactive flapping of its tail. I closed my eyes so I wouldn't see the blood.

I brought the *Times* outside with me, and attempted to read it, sitting on half the paper while I tried to fold back the pages of the other half, which threatened to flap out of control in the wind. The *Times* was the one element of my previous life that I continued to hang on to, at considerable expense, for over a year after I settled in British Columbia. I thought it would arrive by mail day by day, but it never did. It came in batches of four or five issues, and I began dreading to go to the post office, because once I had newspapers I had purchased with so much effort in my hands I felt compelled to read them from first page to last. This sometimes took two or three days, and was hard work. When my subscription ran out, I was so relieved that I never renewed it, but until then I took the *Times* in bundles on the boat, and read an issue a day, throwing it overboard, sheet by sheet, as I finished it. It amused me to look back at the string of *Times* pages floating in a neat chain in our wake as far as I could see.

On this particular day, I soon gave up trying to read. I felt warm enough to take off my sweater and noticed that the wind and groundswell had fallen away. "Do you need me?" I asked, and John shook his head, happily immersed in cleaning fish. "I think I'll take a bath now," I said.

Back in the pilothouse, I shut the door, heated water in the kettle while I was undressing, pulled out the shallow "bathtub," filled it with warm water, and set it down on the *Times* I had spread along the narrow corridor of floor. I anchored a saucepan of warm soapy water in the sink, and began the tricky process of wash-and-rinse from my head to my feet, leaning alternately against the cupboards and the bunk with the boat's roll until I was washing my legs and feet, when I did a one-legged dance. The sun coming in the windows was warming, and my view was of mountains gaudy with snow. The pilothouse had been miraculously transformed into a snug bathroom, and I luxuriated in talcum powder and clean clothes. I put the "bathtub" back under the stove, flung open the door, shouted at John, "I've *never* felt so clean!" and threw the paper in the sea. Bathing every second day or so in reasonably calm weather was a delight. John was so tired at night that he usually saved his bath for a bad-weather harbor day.

John waved to me to come out and see that we were running through a frenzied patch of herring: herring bubbling in the water below; herring boiling up to the surface, asking to be caught. We had two nice fat ones for lunch, on toast. "God, I wish I had my own homemade bread here," John said. "It's got everything in it and makes wonderful toast."

"It's funny. I don't know any Americans who make bread, and I don't know any Canadians who don't," I said.

Another troller had joined our fishing circle, and now we were six, moving slowly in a line that turned, one by one, back in the opposite direction. I sat beside the gurdies after lunch, while John continued to work his lines. Watching a vessel two ahead of us that seemed dangerously close to another troller on the turn, I asked, "When boats are trolling together like this, do they ever hit one another?"

"Once, on the West Coast I had a fellow's bow poles pass over my bow," John replied. "When fishermen say the West Coast, they mean the west coast of Vancouver Island. I paid for my house and educated my boys by fishing there, but it's rough, tough fishing and I don't want to work that hard anymore. Another time, there were thirty-five boats fishing here at the Goslings, and one by one they left, until there were only two of us. This seven-year-old boy was supposed to be watching for his dad, and with the other boats gone, his dad and I were out in our sterns cleaning fish, not paying attention to where we were going, so we were to blame. The other guy broke off his bow pole when he ran into me, but I managed to remove most of his gear and save his pig as it went across my stern. The kid went into his bunk and cried, and I really felt sorry for him. Then there were these two old pals who were fishing in separate boats off Spider Island, just across from here, who ran into one another. When their bow poles locked, they both ran into the bow and shook their fists at one another, and it took about a month before they spoke again."

It was a terrific morning for fishing. I lost count of the times John swung fish into the boat: violet, pink, and gold luminescent flashes from the dripping fish that glistened in the sunlight. He would lean over the gunwale to gaff them, and lift them into the boat. Then always, the same quick action: he would hold the line hooked through the mouth of the fish in his left hand, and maneuvering the gaff hook in his right, with a snap of his right wrist toward his body, pry out the fishhook from its victim's jaw while he was pulling the line away with his left hand, and drop the fish, already turning a silvery blue, into one of the compartments in the fish box. He would check the hook he had just disengaged from the fish: if the hook was bent, he would replace it; if not, he would give the point a quick sharpening with the steel file he always kept beside the gear, then unclip the leader from the mainline to which it was attached, coil it up, and lay it all neatly away. His hand moved instantly to the gurdy to bring in the next leader and hook—all done so fast and with so little waste effort that, unless you watched very closely, it seemed a single motion.

I asked him how many fish he had caught so far that morning. "Ten springs, all big; thirty coho; and a sockeye. The rest were junk," he replied.

John's definition of "junk" encompassed everything he picked up on his lines except salmon or halibut, and included the despised humpbacks. Peter Spencer told me that once when they were fishing and ran into what Peter referred to as "a bunch of pinks" or humpbacks, John said, "That's fine. I'll go in the cabin and steer. You can fish." Then they went directly to the fish camp at Spider Island to a cash buyer, because he didn't want those fish on the boat. "Junk" on our lines ranged from a delectable lingcod to a hideous brown three-foot-long dogfish; or bottom fish like hake or an occasional sole, or the spiny but delicious little rock cod; or seaweed, preferably kelp, which he saved and took home for the garden; plastic bags; pieces of rope and fishnet; jellyfish; and, once, a drowned duck, looking tiny and pathetic. Another time, he caught a surprised seagull, which bit his finger while he was disengaging it from the hook, before it flew away.

Now John put the lines with empty hooks back in the sea, then turned the hose on the checkers of the fish box. He was hosing seawater over the catch and cleaning the bloody sides of the box when the *MoreKelp* half rose in the air and slid back down the side of a great gray wave. I scrambled to my feet and saw black fins—sharp-edged black triangles—cutting through the water between our boat and the one in front of us. The black fins swerved to port, disappeared, resurfaced on the starboard side, and went under. Shortly after, two parallel glistening massive black backs rose in a graceful arc above the sea, diving back through the waves they had made by their hugeness, and were gone. The sea was clear again.

"My God! What was that?" I cried.

"Killer whales," John said. He was pulling in the lines and putting them away. "That's the end of today's fishing."

The fisherman ahead of us was hauling in his lines as fast as he could, too. Where shortly before there had been an easygoing movement of figures between pilothouses and fishing lines in the sterns of boats ahead of and behind us, all were now preparing to leave.

Thirty minutes later, we were running south and there was not a boat left where we had fished. "We're going to Pearl Rocks and the Danger Shoals," John shouted at the brown figure on the *Janice T.* as we came abreast. John Chambers waved, shouting back, "See you in Hardy."

I went in and found the old encyclopedia that John kept under his mattress in his "library" section at the top of his bunk, and read:

> KILLER WHALES: weighing up to nine tons, measuring up to thirty-two feet in length. Salmon eaters. In British Columbia = Orcinus Orca. Among fastest creatures in ocean, capable of speeds up to 30 miles per hour. Travel in pods. Sea-going mammals.

"Sometimes they roll right beside the boat, too damned close for comfort," John said when I rejoined him outside. "It's a son-of-a-bitch when they hook into the gear, because they get scared and sound—dive to the bottom—and our lines are six-hundred-pound test, so you're likely to break a pole and certain to lose the line and a lot of time. Years ago, I heard a story of a killer whale who got caught somehow on a bow pole of a fish boat with the gurdies running, off Bull Harbour. The whale sounded, and when it went down it broke the block off the gurdy. The fisherman jumped into the cockpit, lay flat, and the whale going down gave the boat such a hell of a lurch that the steel mainline cut into the gunwale like a hacksaw. That fisherman was lucky. The whale could have taken a smack at him with its tail if it had gotten really annoyed. It could bust in the side of one of these little boats, I guess, if it was scared enough."

When John had put away all the leaders and hosed down the last of the new fish—two coho and, off the bottom hook, a red snapper—he stepped across the fish box, pulled up the poles, came behind me into the pilothouse, and shut the door. He handed me two substantial snapper fillets and said, "What about a red-snapper stew?"

"Have you got a recipe?" I asked.

"Recipe, woman? Just use your head and whatever you find right here! I'll bet it will be dee-lishus!" He settled down in the seat at the wheel, pushed the throttle forward, and we began running at full speed. "Who uses a recipe?"

"I do," I said. "But you've challenged me."

"Good. Put in lots of garlic," he said, pulling out the logbook and starting to enter the day's catch. "Make it a big stew, because we'll be going to Port Hardy soon and we'll want to have enough for dear Chris Sondrup when we tie up at his dock. He'll be looking for us. We're lucky to have Chris and his beautiful garden right across from the cannery. The horrible brilliant mercury-vapor lights on the public dock at Port Hardy are *very* bad for sleep, and if you tie up at the cannery in the evening to deliver in the early morning the bloody ice-making machine roars all night."

Until now, I had felt too insecure to cook the fish we ate. I made our big breakfast, and the salads and vegetables for our main meal, between four and five in the afternoon. But cooking fish was John's job. Anything new I attempted on the boat, even cooking, it seemed, had to be initiated by climbing a mountain of fright inside myself. I had a moment of panic; my mind went blank. I couldn't remember how to cook at all.

I went out and removed the hatch cover, pulled up the grocery box, and examined the contents. I returned to the cabin with a parsnip, green pepper, onions, celery, carrots, parsley, scallions, leeks, and a small head of romaine lettuce, holding them with both hands in the tails of my shirt, as I unsteadily traversed the distance to the cabin door, with the wind blowing me along the deck of the pitching vessel.

I felt considerably cheered as I dumped my groceries on the counter by the sink, lit the stove, took John's fish knife out of the cupboard door, pulled over a small chopping board that John had made to fit across the sink, and began cutting up vegetables. When the top of the stove felt warm, I put several teaspoons of safflower oil into the bottom of our red stewpot from a bottle among the condiments packed tightly in the curve of the kitchen counter, chopped four cloves of garlic fine and put them into the oil, and set the pot on the hot spot on the stove. Gradually, as I chopped, I added the other vegetables to the pot, beginning with the onions and including the lettuce. What next? I went below into the bow, took the cover off the bench, and surveyed our canned goods. The only red-snapper stew I had ever had was in a restaurant in Baltimore, Maryland, and it had been a brownish red. Tomatoes! I climbed back into the pilothouse carrying two cans of tomatoes, which I opened, pouring the contents into the pot. Then I pulled down the table, and from the spice shelf behind took out basil, tarragon, vegetable salt, pepper, and paprika, and added a dash of each to the stew. An end of a lemon, left over from John's favorite morning health drink of hot water, vinegar, honey, and lemon, caught my eye. I cut that up, skin and all, and put it in, too, plus a swig of Tabasco sauce. I stirred everything with a spoon and tasted the results. Something was missing. The fish! I cut the snapper up into small pieces that I dropped into the soup, and as soon as the mixture began to boil, pushed the red pot to the back of the stove, to simmer gently. John let down the stool and I sat beside him. We were traveling at our top running speed, and the wooded shore was flying past. He said, "How about a hot buttered rum?"

I clapped my hand to my head. "Wine!" I exclaimed, and got the bottle of red wine from the cupboard. "It would be better if it were white," I said, "but this one's already open," and poured some of the wine into the pot. John was behind me, with a spoon. He stirred the stew and tasted it. "Marvelous!" he said. "From now on, I'll stick to fishing and you cook."

Which is what we did.

We went back to our observation posts at the wheel and drank hot rum while the sun slid toward the horizon and the moon appeared, white and misty, in a gray sky. We were running down Queen Charlotte Sound, on the main water highway. We were about halfway between Prince Rupert and the Queen Charlotte Islands to the north, and the ports of Vancouver and Victoria at the southern end of British Columbia—an area that John always sarcastically referred to as "Lower Funland." We passed a green tug going north, pulling a massive barge on which I counted ten mobile homes, half cream-colored with green trim, half white with green trim, on top of some sort of a platform; and then two giant log barges, moving south, very low in the water under the enormous weight of the stripped trunks of great trees. "Those are the huge self-dumpers the logging companies use now," John said. "That load is three million board feet at least. The barges dump their load of logs and are away again in two hours. They just open all the valves on one side, to fill and dump, with practically no spillage. These are coming from the Queen Charlotte Islands." He looked through the glasses again and handed them to me, pointing to what appeared to be a dot on the water. "There is another barge going back to the Queen Charlottes right now." When I looked through the glasses, the dot materialized into an empty barge moving lightly over the waves, north, in the opposite direction.

Miles away but highly visible, a large blue passenger ship whose deck lights twinkled in layers in the dusk of the fading day passed us heading north. John studied it with the glasses. "I can't make out the name, but it looks like the *Wickersham*. I believe that's the biggest ferry the Americans have got on the Seattle–Alaska run."

Three trollers running close together hurried past, several miles away, going up the coast, their poles in vertical running position, with American flags flying on guy wires between their masts and booms. "Look at them go!" John exclaimed. "They are American trollers on their way to Alaska, and by law they must keep their poles up all the time when they are moving through Canadian waters. It's the same for us in American waters."

"God, how your U.S.A. fishermen can run!" John said. "They will start from Seattle with two on a boat at most and *many alone*, and run hundreds of miles, to Sitka, Petersburg, Ketchikan, or some other place in Alaska with only two or three one-hour stops in a sixty-five- to eighty-five-hour run, and then down again in the fall, and occasionally during the summer, too. How do they stand it? Not with flesh and blood! It must be TV and potato chips!"

We were moving with the tide at a steady seven knots an hour, the motor a low roar beneath us. John turned on the Mickey Mouse radio, and a man's voice said, "I have found where the canaries are—lots of canaries." This remark was greeted with heavy laughter by a voice that John said was not too far away. The first speaker went on, "I'll tell you when I see you," and the man who had laughed said, "Roger." Then silence.

"They are using a private code to talk about finding fish," John explained.

Static from the radio, then a different voice: "Romeo, Juliet, and tango." A second voice said, "I can hardly read you, Larry," and the first voice repeated its message. The second one said, "I can't read you at all. Can you slow down a bit?" Silence. "How are you making out down there?" the first voice asked, adding, "I can't pick you up at all now. Give me a call at suppertime and see if things are any better. O.K.?" There was no answer.

John said, "Once, when I was in Freeman Pass, just below Prince Rupert, on a brilliant, clear, moonlit night, it was blowing hard, a northwesterly, and I went into a small bay to anchor. I woke up about 2 a.m., wondering what the weather was like outside. I tuned in the radio and listened to the gillnetters out there. Two Kitkatla Indians were talking, and one said, 'How you doing?' and the other said, 'No good. No good.' So the first one said, 'Well, I got a pretty good catch out here. Pretty good.' So the other asked where he was, and he replied, 'There are lots of boats here, you know.' They shifted to their own language for quite a while, but the second guy just couldn't pick it up, so he asked again where the speaker was. Finally, this first guy said in English, 'Well, I tell you. I am way, way out. *Right* under the moon.' "

Another voice said, "Yeah. I don't know what to do now. I don't know whether I'll get any more fish or not."

John laughed. "It must have been about ten years ago on this fisherman's band that I heard Bull Harbour, the weather and radio station for this middle fishing area, call a Norwegian fisherman known as 'Hungry Ole' by the other fishermen, because of his reputation for getting as much fish as possible and to hell with anything or anybody else, as long as he made the dollars. Ole answered on the radio, and the operator at Bull Harbour, sounding upset, said he had a telegram for him. He read the wire slowly over the radiotelephone, which was open, of course, and the message was that Ole's wife had died. When the operator got through reading, there was a short silence. Then the voice of Hungry Ole, with its heavy accent, was heard saying, "Yah!

Thank you very much, but *wot* a time to die! Right in the middle of the halibut season!"

"I don't believe it!" I said.

"It's true," John said. "I heard it myself."

A new voice interrupted. "B.C. Packers hasn't come to collect my fish," it said, and a second man replied, "What's wrong with a cash buyer?" To which the first answered, "He offered me fifty cents less a pound for coho. I'm not going to sell to him at that price."

"I don't understand about collectors, packers, and cash buyers," I said.

"Most of us in a given area sell our fish to one company," John replied. "For many years, when I was fishing north around the Queen Charlotte Islands, I delivered my fish to the Prince Rupert Fishermen's Cooperative, in which I still hold shares. Now, during the peak of the season, I fish entirely in this middle area and deliver to one of the few really independent companies left, Seafood Products, in Port Hardy, on Vancouver Island; except when I fish on the other side of Queen Charlotte Sound, also in this middle area, when I go in to Namu, which is a summer operation run by B.C. Packers, a part of that bloody big conglomerate monopoly Garfield Weston. Trollers like me dress our fish and put it down on ice and deliver it to the canneries ourselves—usually every six to ten days. Trollers with freezer boats deliver only when their freezers are full or when their refrigeration unit breaks down, which happens all too often. Net fishermen—on the small gillnet boats or the big seiners—sell their fish in the round (that's without being gutted), so they can't hold it for more than a few hours, and the collectors from the companies come on the fishing grounds every day to gather up the fish. They book it in and settle with the fishermen later. That way, companies save hundreds of thousands of dollars in interest on the money they would have to borrow from the banks to pay the fishermen outright."

John swerved to avoid a rotten piece of wood that was bobbing along on the surface, and then a long log that three gulls were sitting on, picking at each other's feathers, and continued, "Packers are vessels built to transport fish. Cash buyers are men who come around on packers ready to pay cash on the grounds. Some fellows sell *all* their fish to cash buyers, but many use the net racks and other facilities of a company or charge fuel and supplies during the season to the company store in the fish camps that companies maintain in places like

Namu or Smith Inlet, which is near here, or on the west coast of Vancouver Island; and the fishermen have to sell enough to the company to cover all that. It's a great temptation for some fishermen to sell to cash buyers, who don't have anything but a boat and weighing scales and then are off to the Campbell Avenue fish dock in Vancouver. B.C. Packers used to have spotter boats out, and if they caught guys who owed them money selling to cash buyers they would foreclose on their boats. Some of our Indian fishermen never saw cash at all when B.C. Packers had its own fleet of fishing vessels. Of course, when it's foggy, the cash buyers had a field day and the spotters didn't have a chance!"

He reached up and snapped off the radio, which was crackling with static. "What a noise! Fishermen and packers often go deaf, and most of my pals who have worked on tugboats have badly affected hearing, too. The older tugs aren't so bad, because their wooden hulls absorb some of the noise, but the tugs with steel hulls and Jimmy motors—that's the General Motors diesel engines—are terrible."

A few streaks of yellow light remained on the horizon in the twilight, and then it was night; we were soon traveling down a bright path to the moon. John got up and went over to the counter, picking up a bottle from the loosely packed items there, shook several capsules into his hand, and swallowed them with water pumped from the sink. "Vitamin E," he said. "I'm going to give you a bottle of your own for your next birthday, and you will be much better off physically and mentally. I've been taking it since I stopped that blood-thinning medicine that made me so sick after my heart attack in 1963, and look at me now! Here I am at sixty-one, financially O.K., happier than hell with my life, and you will be, too, when I've finished building you up to become part of the B.C. coast. I'm working as hard as ever, from April until the end of September. I'm actually unhappy if I miss any potential trolling time, and it can't be explained anymore by economic pressures. I ask my retired troller pals about this, and some who are seventy-five or eighty years old say they wake up at three, four, and five in summer and are mentally running out with us in the morning, catching fish in their minds. That 1963 heart attack did straighten me up to this extent: I am trying to say to John D. every day, 'Slow down, sleep in occasionally, you are well off, there will be fish tomorrow.' But I'm warning you. We trollers are a queer lot."

"Were you fishing when you had your heart attack?" I asked.

John said, "I had my heart attack when we were unloading fish at Prince Rupert. I was lucky. Sean, who was seventeen then, was with me, and Dick, who was twenty and in his last year at university, was working in a cold-storage plant that summer in Prince Rupert. It was the beginning of August, and Sean and I had been fishing up and down the coast, coming in every ten days or so to unload. We would drop in then to see Dick, and bring him fresh salmon and homemade bread we baked on the boat, at a house he shared with other students in town. We had been pushing really hard because Sean was going to start university in the fall, which meant he had to be in Vancouver by mid-September. Sean was down in the hold and I was working up on deck when I got this awful pain across my chest. I knew what it was right away, so I called down to Sean to come up and went into the bunk. I told him to call an ambulance, that I had had a heart attack, and the next thing, I was in the hospital, in an oxygen tent. Sean called Dick and met him at the hospital. I was in the hospital six weeks, and the first time I tried to take a walk outside I had to sit down on every curb. Sean stayed on the boat and looked after it, and in a few hours everyone in town knew I had had a heart attack, so there were a lot of concerned friends keeping an eye on everything. I got excellent care, because the doctors in Rupert knew all about ailments affecting fishermen, and hearts are one of them. We get very little sleep and stand in one place all day without walking, using our arms all the time, which is taxing on the heart. I was in an oxygen tent when Dick and Sean came to see me, and Dick said I looked like a Bedouin chieftain. I was a dark tan from the sun and westerly winds we'd been having, with my beard—it was getting gray then—sticking out over the white sheets and that green tent over my head. On their second or maybe third visit, I said to them that it was no use ending the season like this—why didn't they take the boat out and continue fishing? After all, they had both fished with me since they were ten or eleven, so I suggested they go out for one trip locally to get their bearings and then fish the *MoreKelp* back home to Garden Bay. Sean had been monkey-wrenching the engine all summer anyway, and both were experienced fishermen by then."

John yawned, stood up, and shifted the wheel until he was satisfied with his course, leaned down, yanked the rope to put on the automatic steering, and sat down again. "They did quite well on their first trip out of Prince Rupert, with all my old fishing pals watching out for them, and then they headed for home," he went on. "It was a good test for their self-reliance and eliminated my main worry, which was how I was going to get the boat home. I was very pleased. Then, when I came down from Prince Rupert, I went to my aunt's to recuperate, on Vancouver Island, at Cowichan Bay, where I had spent my childhood."

My rum drink was cold, so I added hot water from the kettle. "How would I get help if we were fishing and I needed it?" I asked, as casually as I could.

John took the steering off the Mike and swerved to avoid a large clump of kelp and broken timber floating around us in the tide. "You would call out on the radiophone," he said gruffly.

There were four radios on the *MoreKelp*: the little citizens-band Mickey Mouse, at the ceiling, for local listening and conversations; the VHF, a neat square modern box for making telephone calls through the nearest land relay station, with dozens of channels for different purposes and a range of thirty to forty miles; a little AM–FM radio over the bunk, on which we listened to weather reports and ordinary radio programs; and John's favorite, the big, ancient black oblong Wesco that looked as if it contained a homemade bomb, which sat on a shelf on the wall at the foot of John's bunk and had a beautiful deep tone. This was the one he tuned in to the CBC while he was outside fishing; and for the news and weather.

John took a small round microphone off the hook on the side of the VHF radio and held it in his hand. "We aren't supposed to have the old radios anymore, because they are powerful enough to block long-distance channels. The VHF would be your best bet for getting help. You switch it on, get Channel 16, and holler 'Mayday! Mayday! Mayday!' three times. Listen three minutes for a comeback, and if there is no response, repeat it, and say 'Over' to give the signal that they can speak to you. As soon as you stop speaking, take your finger off the button, because otherwise you can't hear anyone answering you. 'Mayday' alerts everybody—your closest boat neighbors, the Coast Guard, Air and Sea Rescue—and when they answer they might ask you to shift to another channel to talk. When someone replies, remember to put your finger on the button to speak, and state your position: where you are. Do you know where you are?"

"No," I said nervously. "I haven't the faintest idea."

"Oh, I think you do if you just stop and figure it out. That's one reason I'm so insistent that you learn to read charts." He pulled down a chart. "Show me where we are."

I hesitantly traced our journey from the Goslings south along the west coast of Calvert Island, the large land mass we had been passing for several hours. I figured we must be close to the southern end of it by now.

"That's right," John said. "Just be sure you take your finger off the button when you aren't speaking."

"I'll have to practice," I said. "I'm not very mechanically oriented. Then what?"

"If the Coast Guard answers, they will generally try to get someone locally to come right over if they can, and others will come, too."

I must remember this, I thought. I must remember all of this, so I can hear him talking in my mind, in case he can't speak to me.

The subject had exhausted us both, and we shifted with relief to a discussion of what we would have for supper, which I began to prepare as John was turning the boat toward shore, into a small harbor marked on the chart: Grief Bay. A black-and-silver harbor; silver where the moon's rays fell on the water and the rocks, bordered by trees shading from gray to black. When I got up at two in the morning to use the pail on deck, I saw a cluster of twinkling lights moving on the dark horizon outside the harbor; and quite near us inside, a single light rocking slowly at the top of a shadowy mast on a shadowy boat that proved in the morning to be a tiny troller, the *Silver Dawn*, which had slipped in during the night. John told me the mass of twinkling lights I had seen was probably from the Canadian ferry boat *Queen of Prince Rupert*, on its regular northern run from Port Hardy, on Vancouver Island, to Prince Rupert.

VII

The sun was rising as we ran out of Grief Bay into Queen Charlotte Sound: a gorgeous sunrise, a white day. White sky, a bright gold sun shining on blue-gray water, and shadowed mountains. Signs of feed were everywhere. Little birds—dipchicks, John called them—bobbing in pairs beside the boat as we sped along, ducking their heads under the water, their feathered rears bouncing above the surface. Gulls flew overhead, although we had no fish yet to offer them: gulls out scouting for an early-morning breakfast around and above us, feet drawn flat against their bodies to create a smooth flying surface. Through the glasses I watched one come down toward the deck; it banked, holding its lower wing stiff and floating to the side, wings beating ever so gently. I broke off a piece of toast and threw it out to a group of gulls riding the waves alongside us. Immediately, all the birds were circling and swooping. A gull neatly picked it up in his beak, taking his prize off into the cold blue air, pursued by the others; all circling higher and higher until they finally moved off in a straight line in the same direction as we were going, to the fishing grounds.

We passed three gulls standing on a floating log. "Do you think they have been standing there all night?" I asked.

"Oh, yes," John replied. "They look for food, but they go back to the log to rest, and they sleep there. I always wonder if their legs tire as much from standing as mine do. See how they sort of bend their legs in the swell and turn around and talk to one another! And how they are always scratching and picking. That's because they have a tremendous lot of lice."

The weather report said "Sea rippled with a moderate swell," and the sea slowly billowed in majestic surges. Out of the pilothouse window, I counted six gillnet boats that had shown up ahead of us, four in single file, with two stragglers. In addition to the drums on their stern decks, they all had two main trolling poles as well, and were what John called "bloody combination boats"; they could troll when the gillnet fishery was closed. John had gone to the stern and was putting out the gear. The groundswell had become so pronounced that the vessel immediately ahead of us disappeared in an enormous trough, only to reappear on the crest of the next wave. I went out to see how John was coping. Without pausing in his rhythm of hand-on-the-gurdy, hand-on-the-line, clip-on-a-leader, back-to-the-gurdy, he said, "See that big one coming up?"

I turned around in time to see that we were going to be swallowed by an enormous gray wave. Instead, we glided over it, rolling gently. "It must be terrible for anyone who gets seasick," I said.

"Awful," John agreed. "I had one of my old ski pals along with me who wanted to see what fishing was like, when the sea did this for days, and not a breath of air where we were, either, although it had been blowing like hell offshore. I don't think he ever wants to see a fishing boat again."

We were passing close to a rock pile jutting out of the water. On the top ledge, a lone bird kept vigil. Looking up, John pushed back his visor and shouted, "Get the glasses! Hurry!" I ran in and got the binoculars. John looked through them at the bird, then passed them on to me, saying, "This is Watch Rock, but it should be named Eagle Watch Rock, because that eagle is almost always there."

I could see the bald eagle's head and wickedly curved beak with my naked eye. While I was looking, it made a leisurely, dignified descent to the water, plucked a fish from the waves, and flew back to its perch, holding the fish in its talons. Watch Rock vanished as our boat dipped down into the trough of a groundswell, and when we were riding the peak again and the rock reappeared, the eagle was still there. It was turning its head from side to side, searching for a new victim.

John pulled the lines in one by one, full of fish; I counted ten springs, six of them mild-cure, three coho, and, to my surprise, *two* halibut, each about sixteen pounds, in the fish box already. John was swinging another bright-red snapper on the line over his head. "I just got another Norwegian turkey," he said. "I'm glad to see halibut coming back; that's two in a row I've caught. I used to catch four or five a day here. Halibut trolling is a specialized form of fishing; you troll very slowly and keep your hand on the line and bounce the bottom with your gear all the time. It's a good way to lose gear on a rocky bottom. Fifteen or twenty years ago, some of the halibut fishermen were getting up to a hundred and fifty of them a day."

John picked up one of the big springs, made the usual slit from tail to throat, peered in, laughed, and, beckoning to me, said, "How would you like a stomach full of shrimp like that?" I glimpsed a mass of tiny shrimp before he threw the guts over his shoulder, and a gull swooped down and caught them in midair, swallowing them in one gulp while its companions screamed complaints. John washed down the eviscerated salmon, put it in a checker under the burlap, picked up a coho, and continued his cleaning operations. "The Norwegians here pronounce coho as *coo*," he said. "I once heard a Norwegian I knew talking to some friends over the Mickey Mouse, and one of them said, 'Come up here; we're getting lots of coho,' to which he answered, 'You stick with the coo and I'll stick with the salmon.' He was fishing springs, of course."

One of the halibut had suddenly come to life enough to flap its tail against the deck of the fish box. John thumped it on the head again, and this time it lay still. "Halibut and coho get up much later in the day than spring salmon, and have a late lunch or supper," he said. "It's very seldom you get one halibut before two in the afternoon, let alone two of them. They especially bite late on trolling gear. I don't know why. These are small halibut."

"How big do halibut get to be?" I asked.

"I caught a hundred-and-fifty-six-pound halibut off North Island in the Queen Charlottes in the early spring, and I helped another fisherman, Nick Pullman, get a two-hundred-pound halibut aboard his boat, fishing in May off Spider Island, in this middle area north of here," he replied. "I was fishing alone when I got the hundred-and-fifty-six-pounder, so I had to shoot it to bring it on my boat. That west coast of the Queen Charlotte Islands is wild, wild, and it scares me. It's also *very* damp and cold—an ugly, bleak, damp cold."

John continued fishing around Watch Rock for another hour or so, adding six more springs, all over twelve pounds, all mild-cure, to our catch. Suddenly he said, "I can stand just so much fighting around this big rock, so we'll troll our way down to the Pearl Rocks and fish the evening slack there." Where we were, the sea was a deep blue. We came to a place that rippled and looked, by contrast to the surrounding swell, like a road through the sea. We were running below the southern tip of Calvert Island, and on our port side there was a scattering of islands. I located our position on the chart; we were outside Rivers Inlet.

The "road" through the sea seemed to roll on a diagonal, and I said, "Could that be tide?"

"It's the tide, all right, coming up from Fitz Hugh Sound to meet the tide from Rivers Inlet," John said. "Fish like turbulence, so that may be why there are fish at Rivers Inlet, which is usually closed to commercial fishermen, to conserve the spawning stock. Spring salmon swim up the freshwater rivers and creeks, going into a lake system to spawn. Here at Rivers Inlet, the bloody sportfishermen are allowed to fish all but one little place, so they just sit around and wait for the fish, but their boats are so highly commercialized with sounders, radar, loran, and the like that I don't think they should be allowed to fish in any of it, either."

I steered while John cleaned the rest of the fish and put them down on ice. At one o'clock, he was ready to eat and have a rest. He turned off the motor and we drifted, while we ate and listened to his favorite nonmusical CBC program, "The B.C. Gardener." A man with a clipped English accent named Bernard Moore told us how to use cuttings and seeds for replanting in the garden. As soon as it was over, John yawned, went outside, flopped down on his back on the deck beside the hatch, covered his eyes with a sock, and fell fast asleep.

I had been looking wistfully at the miniature oven. I had used it so far only to warm food or dishes, and itched to try some serious baking. John had already told me that to achieve a high enough oven temperature to bake I would have to bring in heat from wherever I could find it, and I had seen him do it once. I shut the pilothouse door and all the windows, stuffed a potholder in the air hole in the ceiling, and pulled the lever on the back of the stove to close the vent and send heat around the oven, instead of losing it up the stovepipe. As an easy start, I decided to make a single-layer cake from a mix we had brought along. There were no bowls on the boat, so I stirred the ingredients in the top of the double boiler, discarding an ancient eggbeater as too rusty in favor of a soup spoon. I found a little square baking pan upended in the back of the cupboard, and poured the batter into it. When I guessed that the oven heat was high enough, I put the pan into the top, or "hot," shelf, closed the door, and waited. Three-quarters of an hour later, John looked in and reminded me that the Music Hour was about to commence. I took the pan out of the oven and laughed. The cake was half an inch high on one side, increasing to two inches in height on the other. I hadn't noticed the *MoreKelp*'s list to one side while we drifted. I took tea and the cake out to John, feeling a little sheepish when I showed it to him. He grinned, cut himself a piece, and said, "This is what fishermen call 'mugging up.' Most fishermen have two main meals a day, and a minor one; and about seven small ones." My first try at baking tasted all right, but I never forgot that lopsided first cake and always turned the pan around every ten minutes when I was baking from then on, even when we were tied up in port and it wasn't necessary.

We demolished that cake while selections from Puccini's *La Rondine* and Verdi's *I Lombardi* rolled out of the loudspeaker across the deck, voices soaring over the flapping of the guy lines, the singing of the wind, the sharp cries of birds. Even now, listening on far better radios to any of the operas I heard when we were fishing, my inner ear adds the lost sounds of the sea to the music.

John selected fresh gear from the collection hanging above his bunk, and while I steered I heard him down below rustling through the boxes of hooks, hoochies, spoons, plugs, and flashers under my bunk. I asked what lures he had used to catch all those fish that morning. "I caught the big springs on plugs and used those old reliables, brass spoons, too," he said. "I got one on a plain yellow hoochy that I picked up last night before we went to bed. I seem to be getting springs on coho gear. I don't know why."

He took over the steering at Pearl Rocks and asked me to come out and watch for logs and kelp. He was changing the lures on his lines and leaning down to turn the wheel in the cockpit, doing several things at one time, since he was also looking ahead periodically. I reminded myself that before I came he *had* to do everything, including the cooking. We were coming up fairly close to one of the biggest of the Pearl Rocks, a bare little island with jagged edges. He slowed down and handed me the binoculars. "Have a look at those sea lions," he said. "The bull is on that farthest ledge that juts out. The females are lying around on the ledges behind and above him."

I watched spellbound as the bull, an enormous creature that had positioned itself facing us in front of all the others, swayed from side to side, confronting us from its perch and roaring fiercely. We swung in closer, and at some sort of signal all the females dove into the sea. John slowed down so that the Gardner barely murmured: wumpety-wumpety. We moved gradually away, and the females climbed back onto the rocky ledges below the swaying bull, which with lifted head and open mouth was presumably bellowing; but the wind carried the sound in the other direction.

John leaned over the side of the boat, straightened up, and said, "I thought I heard the whoosh of a sea lion. God, I hope not. They never stop chasing you once they start, and when the damned things are too friendly they blow their stinking breath all over you. That's bad, but there is *nothing* worse for that than a whale! I don't understand why. My guess is that it's because their air passage is coming out of their guts and all kinds of things get stuck in it. I had a whale beside me at Cape Scott once. I heard it blow twice in the fog and never did see it, but holy God, it certainly did stink! Whales must have gallons of air, because it just about knocked me out. Anyway, we're lucky today. The wind's blowing in the opposite direction."

We moved slowly away at trolling speed. After a while, I went inside to read. I had been reading for perhaps a quarter of an hour when the door slid open noisily, and John was standing there, shouting, "Come out, will you? I've lost a pig in this horrible bloody groundswell, and I want you to keep an eye on it while I turn around," and rushed out again.

I ran after him. He was slowly turning the boat in a wide circle, aiming back at the pig, which he had dropped in the water while he was removing it to bring in that line. The white rectangle of Styrofoam was serenely floating along, not too far away. He turned the wheel over to me and leaned as far over the stern as he dared, gaff poised to strike. "Don't take your eyes off that pig!" he exclaimed. I steered us slowly past it, getting as close as I could, and with one long swoop of his right arm he hooked the gaff to the raised clamp on the pig and lifted it into the boat. No time for cheering; the pig lines were a mess. "Too shallow here for my pigs," he muttered. "I should have known better."

He had me keep cautiously turning until we were headed again in the right direction. Then he set a box down beside him in the cockpit for me to stand on and hold the lines apart while he started to untangle them. "How did these lines ever get so mixed up?" I asked.

"When I dropped the pig in the water and tried to catch it, I hit the handle of the starboard gurdy with my body and the gurdy backlashed and tangled the heavy wire on the reel," John said. "I lost a lead and some gear: a flasher and two lures—all in all, quite a few dollars. I'll try to make up for it in the evening slack at the Danger Shoals."

Shortly afterwards, I descended into the hold onto the ice, which had melted too far down now for me to reach from above, dug out the last of some meat we had brought for this trip, and made John a steak dinner.

As soon as he had eaten, John went right back to work, steering toward the reefs and ledges at the Danger Shoals. "Come outside and watch for logs and kelp," he said. He had started fishing again, pulling in and setting out lines with a kind of furious intensity. I stood in my usual place on the other side of the fish box, looking ahead—at my back, the whir of the gurdies, followed by a pause, a thump, the slip-slop of a fish tail smacking against the sides of the box. I turned half around, leaning against the chain a little but holding tight to the davit for support, so that I could look forward and back at John, too. He was squinting, his spectacles spattered with salt spray. He had finished clearing the port lines, which had brought in three fair-sized springs, and he reset the lines and returned them to the sea again. He bent to turn the wheel with one hand and reached into the cockpit for a bottle of beer with the other. He popped off the top with the opener he was holding, offered the bottle to me, and, when I shook my head, leaned back and took a long swig. When he straightened up and ran his finger around the neck of his bright blue T-shirt, which was stained with splattered blood, he was smiling and his eyes sparkled. A quick glance at the tips of the poles to check on the springs, another backwards to make sure the pigs were lined up properly, running parallel behind the white foam stirred up by the propeller in our wake. He sat at ease again, silhouetted against the sky, which was gray now, with white clouds.

I was thinking of my father, a reluctant city dweller, whose enthusiasm for outdoor life had taken us all to the country for every minute he could spare. "My father would have loved all this," I said. "I sometimes feel I am living an extension of his life. He would have enjoyed being here so much." Thinking about my father, I said, "What was your family like?"

John stood up and threw his empty beer bottle to me. I put it in with a pack of other beer bottles wedged in a wooden slot beside the hand pump. He began pulling the starboard bowline. "The family was English. My mother's maiden name was Broadbent," he said, "and I was named after an uncle, John Broadbent, who was taller than I am and got killed in one of the wars—maybe the Boer War. He was so tall I guess he was an easy target." John unclipped a leader with an empty hook, and dropped the hook end into the pail with preservative. "We both may have been named after a great-uncle, Sir John Fowler, who was an engineer and one of the builders of the Firth of Forth Bridge in Scotland and a pioneer designer and supervisor of the London Underground. 'Heywood' is from my father's family, who were Manx—from the Isle of Man. My father was Heywood Daly. One of our ancestors, Peter Heywood, a lad of sixteen, was a member of the crew led by Fletcher Christian, which mutinied against their captain, Lieutenant William Bligh, on the *Bounty*." He paused to lean over and haul in another big spring, then continued, "Peter was down below asleep when the mutineers took the boat over. When the crew went into Tahiti or Samoa to get women, they gave him a choice of remaining with them or going ashore, so he decided to stay with his naval career and go ashore." He stopped talking. We were passing alarmingly close to some rocks. "This is a very hairy place," he said.

John turned the wheel suddenly. The main starboard line bounced up in the air and sank under the sea again. "Wow! This s.o.b. dropping tide tries to push you over the top of the rocks and you can lose all your gear," John said. "It almost happened again just now. The sounder went down out of sight."

I looked over the side. Menacing gray rocks directly below us, so close, so clearly visible I felt that if I reached down I could touch them. "We just went to eight fathoms and our gear is set for fifteen to sixteen fathoms on the bow poles and twenty-two on the main," John shouted hoarsely above the rising wind, which whistled through the rigging. "See those breakers there? That's a spur that sticks out, and if I could just go over it I believe I would get a few big springs." He was turning the wheel, heading into the rocks again, with an eager confidence. He continued his narration. "Peter was court-martialed in England, and his case went on for three years," he said, glancing up at the tops of the poles for a hint of a bite. "There was a trial, and Peter was finally acquitted because his wonderful sister, Nessie, lobbied so hard for him. She had T.B., as most people did—my dad did, too.

The three or four other sailors tried with Peter were all adults. I think one was shot and one was hanged. Peter ended up as skipper of some Navy boat before he died. I've got a book called *Nessie Heywood* about his sister; three hundred copies were brought out privately on the Isle of Man. In every letter to her brother in jail, she says how much she loves him, and that's probably what kept him going, but that trial finished her off. She lasted only five or six years after that."

John looked over the side. "This tide circles like electricity here, and everything else circles around it. You've got to watch this rock run to Egg Island. You have to get on the right lineup."

"The right lineup?" I said.

"We all used to do it in the old days before loran, but I still use landmarks to help me locate spots where I know from past experience that the fishing is good. For example, I will fish in toward certain rocks, or toward the line of a mountaintop, or a V between the hills, lining up right with it, and moving in that direction until I might see a certain tree suddenly appear behind an island, which I would remember as a signal that the ground under the boat would be rocky there and full of reefs. This would mean that I must turn and go in the opposite direction for as long as, or longer than, I went in the previous direction, depending on the tide and wind; and if I saw a fog on the horizon in one of those hairy spots I would get out of that place as fast as I could." He looked at his watch. "We're still going to fish a bit more of the slack tide, so before we go in I'm going to put us on a lineup with those two peaks over there and the two small humps in front of it at Cape Caution."

We were moving now from the Danger Shoals toward the coast, and where he was pointing I saw two peaks behind two rounded lavender hills, at some distance behind a point of land that I later identified on the chart as Cape Caution. "I've got all my lines out now, and you can steer while I clean and ice the rest of these fish," he said, climbing over the box and going toward the pilothouse. I followed him in, and as soon as I arrived he relinquished the wheel to me. "Just head for the peaks, and keep the bow pointing toward the spot between the two humps," he said. He pulled out the pad that was always propped against the binnacle and made a rough drawing on it of the area. He marked two spots at either end of a long line that went toward the shore and instructed me to turn slowly when I came to them. He roughly sketched land conformations, marking the spaces: Smith Inlet, Egg Island, and flat Table Island. The Danger Shoals was behind us. Then he departed.

I kept an eye on the mountains in one direction, the islands in the other, and as I slowly made the turns at either end, I prayed that I would not tangle the lines by bringing the boat around too sharply. I kept turning my head nervously to peer through the door at the pig lines—even leaned halfway out the window to look back beyond the stern to make sure they had not become entangled.

After an hour, there was a familiar sharp tug on the wheel which meant that John was back in the cockpit, in charge of steering again. I went out to join him while he pulled in lines, winding the leaders up and putting them away neatly for the night, with a new pile of salmon in the fish box. He looked at me with a tired smile, rubbing his hands. "It's very bad to keep hitting your hands on the gurdy; it really hurts tonight." He looked around. "I think everybody's gone," he said. "Very few fishermen stay out as long as I do."

"Why do you stay out so long?" I asked.

He pushed back his visor. "I suppose because I like it. I love it, as a matter of fact. The romance of the thing is so thrilling. The salmon start from six or seven hundred miles up the McGregor River or Skeena River, wandering as fingerlings all the way down through log booms and pulp mills and booming grounds and pollution in the Fraser River until they finally make it to the Gulf of Georgia, out among all the dogfish and seals and every other kind of predator, into the Queen Charlotte Sound; and then they turn around and go back through the trollers, sea lions, seals, gillnetters and sewers, and two hundred thousand sportfishermen in the Gulf of Georgia, up through all the pollution of the Fraser River, into the hundreds of creeks on this coast, to spawn. Think of it! And I've got to outguess the depths they feed at, and when to speed up and slow down, the color and shape of the lures, the size, type, and weight of the hooks, and the color of the water. There are some shades of blue where you seem to find coho, but spring salmon are usually in brownish-color water. I have to remember patterns of runs of various previous years. What I don't know is what's going on under the water." He interrupted himself to nod at a log that was floating by us at a fair distance, with one long-necked cormorant riding on it. "Look, there's a one-seater," he said. "I guess he's a grouchy only child, like me."

John abruptly turned to give his full attention to the line he was pulling in. He unclipped the nearest leader from the mainline, playing his end with subtle grace, as if he were operating a marionette on a single string. I saw a tail flash above the water, creating a great stir of foam, and spied silver glints beneath the surface, which indicated the presence of a very large salmon. The gyrations of the salmon, which was fighting for its life, became more and more frenzied. John stepped back and forth, one minute playing the line off the stern, and the next, as the big fish swung around, moving along the starboard side. John was leaning far out now, grinding his teeth with the effort of holding on to his end of that colorless lifeline of monofilament. Then, as his left hand gripped the line, he reached down with all his strength to bring the salmon's head above water. With a swoop of his long arm, he cracked the frantic salmon on the head with the gaff and, with lightning speed, hooked it into the salmon. Grunting with the effort of lifting the huge fish, he dropped it into the fish box, where its head and tail spilled over both ends. I was dazzled by the iridescent pinks, turquoise, and greens that shimmered among the scales of its powerful long body—as if John had dropped a rainbow onto the shining white boards of the fish box. As I watched, life ebbed away, and minutes later the elegant, glowing creature, which had overflowed its bounds and overwhelmed my senses, was transformed into the lusterless, lifeless fish that was rapidly becoming rigid. I looked back and forth from that aristocrat of the salmon family to John, with awe. "What a battle you two put up!" I said.

"I specialize in spring salmon," John replied. "The rest is incidental."

John was holding the spring down with his left hand while he pushed his glasses back on his long nose with his right. He picked up his cleaning knife and whistled. "This fellow's been caught before, but he got away last time." A bright yellow-and-white hoochy and hook attached to a broken line of filament was hanging from the big spring's mouth alongside John's green-and-yellow hoochy and stainless-steel hook, which were still attached to the leader line coiled now around John's wrist. "See what I mean by gluttony?" he said, and with a twist of his pliers he removed his own hook and hoochy, and then that of the unknown unlucky fisherman, which he dangled a minute in his hand before adding it to his supply.

It was after six. The afternoon had become gray and cloudy, but now, in early evening, the sun was shining in a clear blue sky. John eviscerated the great spring he had caught, making two little ticks with his long knife at the top of the abdominal cavity and gutting the fish. The first thing he removed was the air sac, which, he explained, was a depth-control mechanism that stores the air the gills take out of the water. He threw it overboard, and the long, pink, tubelike bladder floated away on the surface of the sea, attracting a full quorum of seagulls, whose raucous cries filled the air when he threw the rest of the guts over his shoulder.

I looked at the big, round, sad eyes of the lifeless salmon, at his gaping mouth, and said, "I wish he had gotten away."

John said, "Well, I half wanted him to, myself, and he almost did. You have to remember how all of us primary producers of food feel. Everything pyramids on food. Just before you came out, there was a huge pull on the other line, and one of *your* springs *did* get away. Today, I've caught thirty springs, but I only had to weep for twenty-seven, because three springs escaped. They were yours, of course."

On the final line for the day, the first fish he took off the hooks was a tiny grilse, a wriggling, immature spring salmon about six inches long, with clear eyes, glistening silver body, and such a surprised, innocent expression that it made my heart turn over. John handled it with the care and gentleness reserved for a very precious object, holding it behind the gills at the back, with its little head hardly filling the palm of his large left hand, while he slowly, slowly, worked on the hook with his right until—a quick twist—and it was out. He leaned way over the side and gently dropped the little fish back into the water. One final wriggle of its tail, and it was gone. Did I imagine its joy? "I've probably been towing that little bugger for a long time," John said. "Nothing wrong with him except that he has a sore nose now."

We came in past Egg Island, with its big light on a steel tower on the sloping rock bluff and its deep foghorn—a welcome woo-oo sound on foggy days. Egg Island stands out against the horizon; it is wooded, with a distinctive rounded shape that, with a little imagination, can be compared to an egg. It is about ten miles northeast of the Danger Shoals and, because of its prominence and proximity to some of the most dangerous reefs on the coast, acquired a lighthouse station in 1898. We went in behind it to anchor off Table Island, whose top is completely flat—with trees growing out of the flatness like hair. On the way to our night's harbor, we passed a line of gillnet boats across the water from us, anchored side by side; I counted seventeen without the help of glasses, but later, with binoculars, I learned that there were gillnet boats as far as I could see—vessels stationary against the horizon. "They are waiting, with their nets set," John explained. "They'll be gone in the morning. We could fish here tomorrow, but the gillnetters will have cleaned everything out—springs and coho both."

"How many fish do gillnetters get in a set?" I asked.

"I've never gillnetted. Sometimes they get nothing but water, and then again they may get a couple of hundred fish. Apparently, it's just a matter of about fifteen minutes. If you take too long, your net's solid with hake, jellyfish, sharks, and seaweed that make a mess of the net. All junk, although people eat hake and it's used for mink food."

"I wish, when you have time, you would tell me, once and for all, the difference between a troller, a gillnetter, and a seiner," I said.

"Right now is as good a time as any," John said.

As soon as the boat was anchored, John settled down on the steering seat with a sigh of pleasure and his evening drink. He stretched his legs across on the folding stool, and I retreated to the bunk with my pad and pencil.

"A troller, with six or eight lines hanging from his poles, is always in motion," he began. "He does his fishing by daylight, dragging along three to ten lures on a line, or even more, for humps and sockeye. The lures can be flashers with hoochy tails, plugs, spoons, and sometimes bait herring. There are seldom more than two men, or a man and his wife or girlfriend, on a troller, depending on the boat size, which can be anything from thirty-five to sixty feet but averages about forty, like us. Quite a lot of trollers fish alone. The ideal length for a gillnetter is thirty to thirty-six feet, although there are larger ones. Gillnetters over thirty-eight feet in length pull on the net too heavily, and it doesn't function as well. On this boat, for instance, a heavy wind would cause the net to be pulled out of shape. Gillnetters have to move around fast. They fish by setting a net, usually nylon, in the water, rolling it from the drum it has been wrapped around that you see at the stern. The fisherman starts one end overboard, marking that end with an orange Scotchman as a buoy to which a lantern is attached at night, and when the whole net has rolled into the water it hangs down perpendicular, like a sheet or screen. It has a lead line at the bottom to hold it down and floats at the top to keep it buoyant, and the fish simply run into it, catching their gills. The gillnetter has to try and set his net at right angles to the way the fish are running, which he knows from years of experience. To bring in the net, there is a foot pedal controlling the power system to the drum, which he steps on to make the drum rotate. The drum pulls up the net over rollers on the stern, into the boat. He stands between the rollers and the drum, and as the net slowly comes in, he takes his foot off the pedal, it stops the drum, he picks the salmon—and sticks, dogfish, and other junk that come in with them—out of the net, throws the good fish into a side pen on the deck, and the junk overboard."

He shifted in his seat and cleared his throat. "A gillnetter has to be *very* fussy about protuberances; there must be no nails, wooden slivers, or anything else for the net to catch on as it rolls out. All the surfaces it touches have to be smooth, because if the net can possibly catch, it will, even in the fisherman's own clothing, and there are still guys out there wound up in drums and nets. There are variations in net size and color—the smallest mesh are for humps and the biggest for springs—and shades of light green: lighter for river water, and darker for the ocean. A good gillnetter never has less than four nets. And there's a special size for sockeye, and a fall net for the big dog salmon which also catches springs early in the year. Every year, the runs of humps and sockeyes are different, so a good gillnetter's got to remember what he used in the same cycle two or four years before as well. He has to guess at the mesh size when he orders a new net, which costs a couple thousand dollars, and a quarter of an inch difference can mean great success or almost complete failure."

John took a drink and continued, "There have been so many closures for gillnetters that a lot of our Pender Harbour gang have become combination boats, with trolling poles as well, but when they are trolling they must take the gillnet not just off the drum but off the boat. There are usually net racks—elongated sawhorses attached to the floats, where a gillnetter can mend his nets and leave them while he goes trolling—at any place where he sells his fish regularly, but I know a guy who had his net stolen from a net rack at Port Hardy while he was trolling." He stirred the rest of his drink reflectively with his finger. "Most gillnet fishing is pretty well done at night, but the sets at daybreak and dark—I guess you'd call it twilight—are generally the most productive. Timing is critical; ten or fifteen minutes of light, and holy God, if you don't get the set right, if you leave it too long or don't take it out soon enough, it's unbelievable what rises from the bottom: hake and dogfish mostly, and the dogfish can fill up the net and sometimes almost sink the boat."

He got up and opened a can of apricots, went out to the hold and brought in a container of yogurt, poured the apricots into two dishes, and doused them liberally with the yogurt. He laid one dish down with a spoon beside me on the bunk. I was too busy writing to pay attention to mine, but he started hungrily eating his apricots as he continued talking. "Seiners are an entirely different class of boat," he said. "They are forty-five to ninety feet long, with the average around sixty feet, have a crew of four to six men, and can cost anywhere from eighty thousand dollars to over a million. What a seine boat does—it circles a piece of ocean with a net, closing off the bottom with a special line. Whatever is in there when the seine fishermen wind the net in is what they get. A seiner's net is so huge it costs *thousands*, and has a small mesh that is the same for everything but herring."

He paused to remove a pit from an apricot. "When a seine boat starts to fish, first a man is put out in a skiff with the end of the net line. If the seiner is fishing in an open piece of sea, the skiff man throws out a sea anchor—a small canvas parachute that drags—to start the end of the net rolling off the drum. Then he sits and tends the end of the net while the seine boat tows it in a circle to set it. Theoretically, the fish go into the net, and when the ends of the net are brought together the bottom is pursed up the way you would pull together your purse strings. More often, the skiff man takes his end, called the beachline, to the shore, where he ties it to a tree or a post driven into the cracks in rocks on a beach. The seine boat unrolls the net off the drum and tows it in a circle to the skiff man, who brings his beachline out to meet the seiner so they can complete the circle, pull the lines together, and purse up. The critical moment is when the skiff man is untying the line on shore so he can take it to the seine boat. He has to be agile and fast, and you've got the skipper shouting and other boats waiting to make their sets. There's not much left when the seiners get through fishing. They go down from a hundred and twenty to two hundred feet and take up everything."

He had been talking between mouthfuls, and he scraped the sides of the dish. "Seining is so dangerous that seiners don't usually fish at night. You've got an enormously powerful boat towing on a purse line tied to a tree, or post, or stump. Imagine the tension in that line! If the line breaks or the shackle snaps, then the man in the skiff waiting to release the line from the beach is likely to get hit over the head. The shackle hits high on the body, injuring the head, face, or eye, and some skiff men are permanently crippled or deafened when the line breaks. There's big money in seining, even though the government may allow a single set for only fifteen minutes. One seiner I know of made a half-million dollars in one herring set in March off the Queen Charlotte Islands, and I hear of others who've made more—but some years seiners just starve to death. Gillnetters get a steadier income, and sometimes make more money than trollers, but you have to be a darned good fisherman. I don't see well enough at night to be a gillnetter. I can't judge distances the way my neighbor Sonny Reid does."

When the sun had gone down, and our night's harbor was deep blue and glistening black, I went out after John was asleep and sat on the deck. The boats were invisible in the darkness, but their presence was marked by an unbroken line of glittering illumination, endless in the distance. I watched the twinkling lights move up and down in unison across the water with the motion of the waves. Forty-second Street and Broadway, I thought: it looks like New York's theatre district on Saturday night. I marveled that individual fishermen could bring in their sets without cutting each other's nets in half, hemmed in with their boats side by side, so close together.

I wondered which of the shimmering lights belonged to our Pender Harbour fishermen whom I had met last winter. Where would Sonny Reid be in that long line of lights? He had told me he started fishing when he was eight, which meant that although he was only in his forties he had been fishing almost as long as John had. Which boat would be Ray Phillips's, the big, thoughtful logger and gillnetter whom John so admired and with whom he worked hard in the harbor on mutual problems of the UFAWU—the United Fishermen and Allied Workers Union? He would surely be there, with either or both of his sons. What were they doing now? Not sleeping—I was sure of that—on that dazzling thoroughfare of lights between us.

VIII

Before daybreak, I heard the chatter of the alarm going off, the patter of rain on the roof, a groan from John, the sound of feet landing on the floor, and the pilothouse door sliding open and shut. After that, unexpected silence. I slid out of my bunk and looked up through the opening into the pilothouse. John was back in his bunk, asleep.

When he rose, at eight, the gillnet boats across the way had vanished. "Sleepy, sleepy, Lord, I'm sleepy," he said. "I love sleeping with the rain on the roof. Rain makes for better breathing than when it's dry. I have this old friend, Gwyn Gray Hill, an Englishman who has spent his whole life visiting people up and down the coast of British Columbia and Alaska in his sailboat. Gray Hill likes to dance in the pouring rain and shout, 'When it really rains, where are *they*?' By *they* he means tourists, of which there are too many, far too many, in summer at Garden Bay. Tourists are an abomination—one of the reasons I fish this far north."

The rainy day and the warmth of the pilothouse inspired me to cook, since the stove was already on to heat the cabin. At breakfast, after I cut off the end of the bread that was turning green and toasted the rest, I opened a can of mushrooms, browned them a little in oil with onions, and scrambled them with the eggs. "Very good," John said. "I just hope I can work off such a big breakfast in this bad weather." He started the motor, and while we were running out, back to Pearl Rocks and the Danger Shoals, he donned his oilskins—heavy dark-green rubber garments that covered him from head to feet: his hat, a sou'wester, with a broad six-inch brim, which tied under his chin; a heavy coat that buttoned down the front and reached below his knees; and rubber pants that came to the edge of his waterproof boots. By this time, the rain was pelting down. "Steer and watch the sounder," he directed. "Keep it deeper than thirty-five, but don't steer shallower or we'll lose gear." My mind was still on cooking. Reluctantly, I took the wheel. While I steered, I decided that some time during the morning I would make John's favorite dessert—lemon cake. I was trying to write down the recipe from memory with my right hand while I steered with my left, when I felt a chill on the back of my neck, the sharp cold of blunt fright; something was wrong. I should be looking at Pearl Rocks, and what, instead, was in front of me out the window was Watch Rock. Unmistakably, Watch Rock; the eagle was there, on top, slowly turning his head from side to side watching for prey. I had somehow completely turned us around. I hastily began correcting my course, turning the boat in the opposite direction by degrees so as not to twist the lines. When I had us straightened out, I opened the door, cautiously, expecting a blast from John. He was not in the stern, and the lid was off the hatch. He must be down icing fish. If I was lucky, he might not have noticed. I rushed back to the wheel in the pilothouse. A glance at the sounder and I stopped breathing. It had dropped to twenty-two fathoms. Maybe the sounder needle was stuck; otherwise, goodbye gear!

A sudden burst of speed and a firm twist of the wheel by the Unseen Hand; the sounder was showing a safe forty fathoms. I ran to the door, and John shouted from the stern, "Get back in or you'll be soaked. What were you doing?"

"Thinking," I shouted back. "Trying to remember the recipe for lemon cake."

"You had us inside a ledge," John replied. "Lucky for you we got off without losing gear. I'll steer now."

I collected all the ingredients I could remember—margarine, sugar, flour, baking powder, and salt, and then I put on my floater jacket, with the hood up, and ran out in the rain to fetch two eggs down the hatch. Back in the pilothouse, I climbed up on John's bunk and reached across to the top shelf above the sink and got down the double boiler to use as my mixing bowl, and a bread pan for baking the cake. I opened a small can of Pacific Evaporated Milk, measuring out a quarter of a cup of milk to an equal amount of water, and mixed all the ingredients together. At the last minute, I found raisins, added them, poured the cake batter into the pan, and put it on the top rack in the oven. Then I stored the more than half-full milk can out on deck beside the pilothouse door.

Undeterred by high waves that rolled the *MoreKelp* around and roared across the deck through the scuppers, John was fishing. When he leaned over the side to haul up a fish, a stream of water slid down his beaked nose from the wide brim of his sou'wester. When I came out with hot bouillon, he gulped the soup down and handed me his glasses. "They steam up so much I'm better off without them," he said.

I went back inside and turned the pan around in the oven to keep the cake surface level. When John came in for a snack, the aroma of baking lemon cake permeated the pilothouse. He opened the sliding door, creating a pool of water off his dripping oilskins as he stepped inside, then stopped, backed up, leaned down, and picked up the Pacific Milk can outside the door. He held it off in his glove, as if it were unclean, squinting at it without his glasses. "Holy cow!" he exclaimed. "A milk can opened upside down!"

"Oh, I opened that can for my cake," I said. "The milk will keep. I'm going to put it down on the ice when I go outside again."

"Keep, woman!" he shouted, throwing it over the side. "If there were any bluenoses from Nova Scotia on this bloody boat, they would walk right off it if anyone opened a milk can upside down!"

"Are you crazy?" I said indignantly. "I opened that can upside down because the top was rusty."

"Well, I never set much stock on its being such bad luck myself—not really," John said. "But some fellows wouldn't stand for it. I just lost a spring, and that's probably why. He was a big fellow, too."

He began peeling off his gloves. "There was a cook on the *Western Spirit*, a Nelson Brothers Fish Company packer, thirty years ago, who opened a milk can upside down and someone went to sleep steering and they hit the North Arm jetty in the Fraser River and smashed the whole bow. The entire coast knew about it," he continued. "It used to be that if a fellow came on a halibut boat with a black suitcase, they'd chuck the suitcase *and* the man right back on the wharf. They wouldn't possibly allow him in the boat. The worst thing of all," he added, glaring at me, "was a woman on a fishing boat. There was also the superstition that you must never sail on a Friday."

I swallowed hard. "Do you ever sail on Friday?"

"Oh, I think so. Sure. But the halibut fleet didn't, for years."

Black bag, I thought. *I* have a black bag. Don't mention it. "I guess I'm lucky that you haven't thrown *me* overboard yet," I ventured. "Do you really think that you lost that fish because I opened the milk can upside down?"

"I lost two springs in a row," John said. "When I saw that milk can, I thought of that."

"Are you kidding?"

"Not exactly," he said, pouring himself a generous ounce of Scotch and drinking it in one gulp. "I have a couple of friends who would have crossed themselves before throwing the can overboard, even though they never went near a church." He looked at me thoughtfully. "One of those had a wife with a bad case of Stove."

"Stove?" I said. "You make it sound like an illness. Was she sick?"

"That's right. She was always cleaning the top of the oil stove. If she or anyone else happened to get even the smallest spot on the stove, she would rush right over and rub it off and repolish the top. The stove on the boat was bad enough. In her house, in their kitchen, you would have thought you were in a room full of mirrors."

Later on, after I had taken out the cake—baked a beautiful brown—and turned it onto a plate, I went out to tell John it was done, just as he pulled in a nice-looking salmon. "There's one that survived the can of milk," he said, wiping his dripping nose on his oilskin sleeve. "It's a medium-size coho, so we're still all right. The only thing, will the fishing go downhill tomorrow? That's the big question." He was distracted by a log going by us with five gulls on it. "A five-seater! Look at the one sitting in front of the other four, watching ahead. Fishing must be bad, because they're not bothering to get out of their car."

A day or two later, I opened another milk can upside down when I was making a salmon chowder. My first reaction was to cover the label with my hand to make it not have happened. I quickly turned the can right-side-up, as far as the label was concerned, but the opening I had made now faced down. Holding the can in the palm of my hand, with milk dripping through my fingers into the sink, I tore off the label, squished it into a ball, and threw it out the door overboard. By that time, all the milk had leaked out into the sink, so I heaved the empty tin overboard, too, and got out a new can, which I was careful to open correctly.

For the remainder of the day, everything went wrong. Our anchorage that night was, appropriately, Grief Bay. When the orange ball of sun went down behind the trees, the rocks took on a silver light that made the trees and surrounding water seem flat and gray. In bed that night, I checked off on my fingers the number of mishaps since I had opened the two milk cans upside down. Besides my awful careless steering into the shallows and John's loss of springs, I had put wine in a soup, which tasted terrible; forgot to watch for kelp until we ran over a large clump of it that got tangled in the stabilizers; and neglected to refill the kettle that afternoon, so there was no warm water for John to wash in that evening. Worst of all, the stove broke down. One minute, I was watching the dented old kettle rattle around on the stove, with steam rising from its spout, and the next everything was quiet.

I began to feel chilly. I tried to relight the fire in the stove pot, and nothing happened. I went out and gave John the bad news. He was winding up his last line and putting it away, so he came right in. "It'll be bloody cold in here without the boat stove on," he said, as he took off his rubber gloves and handed them to me. I turned the fingers inside out to expose the sopping-wet felt linings, and hung the gloves up to dry while he gathered tools from the shelves behind the vise at the door. "You steer," he said, and sat down on the floor, surrounded by tools, to tinker with the stove. He continued to struggle with it all the way in.

The scenery was changing as we traveled south toward the north end of Vancouver Island. We passed several tugs with mile-long tow-lines to low log barges they were hauling, the barges so far away that to my untrained eye they appeared as floating islands until, with the binoculars, I traced the long line from the barge into the water and out into the stern of the tugboat far ahead. We passed real islands, wooded and rockbound, and other fishing boats; or they passed us, the gillnetters traveling north for what John told me was a new fishery opening for them; the trollers heading south, as we were, toward Port Hardy. We were sheltered by another uninhabited shoreline at our night's anchorage, Port Alexander, where we were the solitary visitors. I had always thought that port meant a city or town with ships coming and going in its harbor, so I looked up Port Alexander in John's copy of the *British Columbia Pilot*, the mariner's bible issued by the Canadian Hydrographic Service, which described it as a "port of easy access, and . . . good anchorage, sheltered from all but southeasterly winds . . . There are, however, no objects to take bearings of when anchoring, the hills on either side being high and densely wooded."

As soon as we anchored, John tried to light the stove. "I've spent endless hours monkey-wrenching this old stove the past month so it would be less temperamental by the time you came," he said. "I've worked on it every spare moment. I've had the exact same setup for years, and it has never stumped me before like this. I think it's mostly an air-lock problem. It was O.K. in harbor calm, but the s.o.b. filled its lines with air when the boat rolled and drove me nuts. At least, it cooked well in harbor. It would eventually always boil stuff, so when it gave me trouble what I did was make stews and soups way ahead of time and eat a piece of toast and half a grapefruit for breakfast; cheese, crackers, and beer for lunch; and then I'd be O.K. until stew time. If I had to fry anything, I'd do that before the wind rose."

He tried several more times to light the stove. Nothing happened. He disappeared below, where I could hear him rummaging around, and he came up shortly with a small box in his hand. "Thank God I've got a spare stove carburetor!" he said. "I try to carry spare parts for *everything*: an extra generator, spare canvas, wood, clean line, fishing gear, and engine parts, including a set of injectors; a full set of batteries and an extra alternator; a spare transmission; even a naphtha-gas cooker, which I keep on the floor here under the wheel in a three-gallon can so that I can cook with the windows open and the wind blowing if I have to. Don't forget that if I don't keep my equipment running well, if my engine stopped in a place like Pearl Rocks or the Danger Shoals, I could smash against the rocks, lose my gear, my boat, and my life."

He removed the new carburetor from its box. "Stove carburetors seem to last three or four years, but I guess the endless motion finally wears them out," he said. "Since you've been on board, the stove has been preparing such loverly meals—pride cometh before a fall—that I've been scared stiff to praise it too much. When I was alone, I got so I talked to the stove and thanked it."

I had curled up on the bunk to watch him, with my feet up out of the way, so he could move freely between the stove and the tool shelves. He knelt on the floor and quickly replaced the old carburetor with the new one. He threw a match into the pot, and the stove turned on. He watched it for a few minutes and then got out the small saucepan and made cocoa. He poured it, steaming, into two mugs, handed me mine, and sat down beside me. "You poor city-insulated types go along flicking switches and turning gas handles without any real involvement until it all quits," he said. "Then you are buggered and one hundred percent bewildered. But as one fisherman who has fought and struggled with machinery since 1935, I am thankful every bloody time that the stove produces a meal and that my engine keeps running, my boat doesn't leak too much, my automatic mike and radiophone keep working. I am not only thankful, I am quite frankly amazed. We battle and monkey-wrench every spare moment to keep it all running, and any time three-quarters of the crap runs at all, *get down and be thankful*. Not to the gods, but be thankful that you don't own more machinery."

That evening, John engaged in the letter writing for which he was famous. Instead of dropping off to sleep at eight, he finished two running letters he had kept going to his sons on the notepad on the compass shelf, tore them off, and sealed them in stamped envelopes, which he took from the lower end of his "office" under the bunk mattress. He wrote several letters on formal blue stationery—one to his Member of Parliament in the Canadian capital, Ottawa, with a carbon copy underneath for his own record, reminding his elected representative of the promise to support a more extensive salmonid-enhancement program (SEP) for the west coast; an order for boat parts from a Vancouver firm to be sent to Port Hardy; and a half-dozen letters to relatives and friends. Observing the fiendish energy with which he attacked his correspondence, I could see how his thoughts, racing so far ahead of his pen that he appeared to be hurling them at the page, produced his demonic and scarcely decipherable handwriting.

IX

Before the stove started working again, the pilothouse had developed enough of a chill to give me a sharp appreciation of how much its well-being added to ours, even in midsummer. I said my own private thanks to the stove early the next morning when I was cleaning it, apologizing for thinking of it as a rusty derelict the first time I had seen it on the marine ways when John was scraping the hull. The stove had not been used then for weeks, and the top had been covered with rust. I had watched later when John cleaned it up, magically producing a smooth, glistening iron surface, but I had found that with use the top deteriorated so rapidly into spotted rust that it always needed cleaning. Now, as I had seen John do—I had written down his cleaning method in my notebook—I turned off the stove and, while it was cooling, swept the top free of crumbs and loose dirt with the whisk broom and dustpan, shook Ajax on a piece of copper screening that I held in place with a damp rag dipped in vinegar, and began to scrub: hard and harder. Each spot had its own stubborn resistance. When the surface seemed clear of spots and free of rust, I ran wax paper, waxed side down, over the warm top, and when the melted wax from the paper didn't seem to bring out enough of a shine, I tried candle wax, and behold—the same kind of gleam that John had achieved.

The waxed top smoked a bit when I turned the stove back on. The pilothouse was filled with the acrid smell—delicious to me—of warm vinegar and melting wax. "Don't burn us up," John said, sniffing, his head in the door. He came in, looked approvingly at my handiwork while I stood proudly aside, and added, "If you have any letters you want to send off, get them ready. We're going to deliver our fish into Port Hardy this morning."

John had once sent me an air-view picture postcard of Port Hardy with the message, "Where we shop, and the only town you see most trips—main drag could *not* be more 'Chamber of Commerce' *horribly* ugly, but it's surrounded by beauty." Until now, Port Hardy had been a postmark to me, stamped on daily messages from John: a whole week's thoughts in one long, running comment stuffed into an envelope, together with scraps of any kind of paper on which random thoughts had been jotted down while he was fishing from the stern or otherwise occupied. Right after we met, his first letter to me, from Garden Bay, had been composed on the circular shipping cardboard around an electric light cord; and another, on a dozen numbered sheets from a tiny pocket notepad, ecstatically describing the view and his emotions, had been written while he leaned against his crossed skis in the snow on top of a mountain.

I steered, while John prepared for our arrival at Port Hardy. He lifted the office end of his mattress again to get a plastic bag stamped "Fresh Vegetables" into which he dropped all our letters that were ready for posting. Then he went down in the hold and brought up several salmon and a small halibut, to give away; and in a separate fish-box section he piled up four beautiful mild-cure-size spring salmon that, he explained, were for a friend named Mary Gunderson. "Whenever I come into Port Hardy," he said, "I bring her salmon to smoke and can for me. I bought her the canning equipment, and she keeps half of what she cans; the other half is for me. Mary is part Indian, from Cape Mudge, on an island along this coast. Nobody else smokes and cans salmon as well as she does." After that, he took the scissors from their hook over the sink, lifted the mirror from its niche against the wall on the counter by the door, and carefully trimmed his beard close to his chin, and then his mustache, grimacing as he snipped the edges.

We changed places, and John steered while I collected the dirty laundry to take ashore in a large black plastic garbage bag. After twelve days, we had quite a bundle. When I finished tying up the bag, I came up to stand beside John and saw that we had rounded a bend into a bay and on our starboard side were passing, first, a red government dock, and then houses. Ahead of us along the beach, beyond the houses, was a group of yellow-and-white buildings that had docks running out into the water.

"That's Peter Spencer's house," John said, pointing to a two-story yellow frame house, with brown gables, sitting in front of a low bluff, a strip of grass between it and a rock wall at the water's edge; and then almost immediately, "There's Lillian O'Connor's beautiful garden," as we passed an imposing white house with a grassy lawn that sloped down to the beach. "Her father-in-law, Terry O'Connor, the father of the three O'Connor boys, was my great fishing pal. He taught me a lot of what I know about fishing. He's dead now, but I still see and talk with him, the way I do with my dad, or old Tom Stanier. Terry had one area toward Cape Scott that he called 'Poverty Flats' where he stayed and fished springs. He was extremely good at it, but I think he carried it a bit far, because he wouldn't fish for anything else. He would make any effort, put on any gear, to catch them. He was a real expert; he fished circles around me." Then, after a pause: "My friends here usually watch for me when it's time to be coming in again. They will have seen us coming by now and will have passed the word around."

We were turning toward the rectangular yellow-and-white buildings. "That's Seafood Products, the cannery where we'll deliver our fish. It's one of the few that are not owned or tied in, in some way, with B.C. Packers. God, what a traffic jam!" John said. "Don't talk to me now. I've got to concentrate. We'll probably be all day getting rid of these fish."

Ahead of us were other trollers with their poles up, like ours—thin, long arms pinned to their sides. Suddenly we were in the middle of the maelstrom: boats traveling alongside, behind, and in front of us, going and coming from and to the same place. The floats in front of the cannery were so chock-full that I couldn't see an empty spot for us, but John was threading his way in, leaning anxiously forward, turning the wheel slowly from side to side, steering with both hands. We crept along, stopping once to back up and let a departing vessel go through; turning quickly to avoid a vessel that was moving toward the oil dock we were passing. At the cannery, John tied the *Morekelp* to another troller, which was tied in turn to a gillnetter, at the far end of a narrow float around the pilings for the cannery buildings above us.

John pushed the clutch to neutral, went outside, and, taking the bow rope in his hand, stepped across to the boat beside us, where he quickly made a knot with our rope on the stranger's bow cleat, then rushed back to fasten our stern rope across to our neighbor's stern cleat.

He returned to pop his head in the door and shout, "Turn off the Gardner!" and left. I watched him walk effortlessly across the other boats, step out on the narrow dock against the wall, momentarily disappear around the corner, and emerge in a broad space at the other end at the steel ramp, which he climbed; and then he was gone. I went back and pulled the green rope that turned off the motor and waited for the buzzer alarm, feeling like a veteran. When all was quiet, I went out on the stern deck and, by craning my neck, saw John standing twenty feet above me, talking with a handsome, smiling young Japanese man. The young man stood on top of the wall beside a large cable winch, directing the complicated scene below.

John disappeared again, so I looked around me at the immense gathering of boats. The hulls were of various colors—mostly blue or green and many black, brown, or white. I saw an orange hull, a yellow, and a red one, and several boats, their hulls made of unpainted silvery aluminum, that looked like baby battleships. Most of the wooden-hulled boats like ours had an aura of old-fashioned grace that, to my biased eye, the more modern fiberglass cabin-cruiser types could not seem to achieve. Right now, all the trollers had two things in common with us: poles pointing skyward, and two racks of idle gurdies in the stern.

Looking up, I could just see a corner of the unloading area—a noisy bustle of machinery and men. Three winches, set on platforms on the wooden dock above, were sending down buckets, fastened by chains from their four corners to the winch cable, into the open hatches of waiting vessels; they came up with brimming loads of fish, tails and heads hanging out over the bucket edges. I had to shade my eyes from the sun. A glance at my watch: eleven o'clock.

John returned, walking casually across the intervening boats. "Somebody at the cannery spotted my poles when we were coming down the bay and, before we got in, put me on the waiting list to unload, so that we might be out of here in a couple of hours. God, I hate the waiting around! I like to unload and get it over with. You'd better take the laundry to the oil dock now. With this crowd, there will probably be a long line at the machines."

He picked up the laundry, which I had brought out on deck with me, and started back across the boats. I followed, stepping carefully across our gunwale while I held on to the rigging of the adjoining boat, crouching on the higher rail that John had gone over with such ease, and letting myself down slowly onto the neighboring deck; crossing its stern, feeling like an intruder as I passed by the open pilothouse door, past some stranger's life, where I could see the remains of a breakfast in the sink; then sitting on the high rail and moving my legs slowly over to yet another level on still another stranger's vessel—this one lower, with a sloping cabin top where there was nothing to hold on to, so I crossed on the roof on my hands and knees and hoped nobody was watching. I dropped down from it, with John's hand to steady me, onto the dock. John laughed as I clung to him for a second. "Wait until there are ten or eleven boats between you and the dock," he said. "I'm breaking you in on an easy one."

John handed me the laundry, and I walked shakily along the narrow float, whose springy action was a reflection of the water churning from all the movement of the boats. I was holding the laundry bag in my left hand and reaching with my right for a creosoted piling under the high dock for support wherever I could find one. Never seasick on the water, now on land I was rocking to the motion of the waves I had just left. I felt extremely dizzy. I was to feel this rocking motion whenever I left the boat after several days of steady fishing, and, once on land, had to hang on to whatever was available to keep from losing my balance.

My courage, always wilting at the unexpected challenges I faced, collapsed at the sight of the ramp I was about to climb. It was almost vertical. I had forgotten how far ramps could drop at low tide, but I had no choice. The ramp, made of rough metal webbing, was divided in the center, with clear passage on the right for freight, and on the left, thin crosspieces, like steps, for people. I started up the clear side, because it looked easier, putting one foot firmly in front of the other, and slipped, so I quickly moved to the crosspiece steps to get a better footing on the steep incline, arriving breathless at the top. I looked down: the oil dock where the small room that housed the washing machines was situated was below me, where I had just been. I had made the mistake of following John up the ramp without thinking. It was worse going back down. I thought, I will surely topple on my head if I let go of this handrail. At the bottom, it took me several minutes to figure out which of three floats led to the oil dock and find my way there through the labyrinth of berthed boats. Several fishermen were lined up ahead of me, waiting for the machines. I set my laundry down at the door, made myself known to the man behind me in the lineup, and wandered around the corner into the marine-supply store that faced the gas, diesel, and stove oil hoses to the pumps that were in a shed next door.

I sniffed inside at the rich convergence of odors: from the gas and oil beside it; the sharp saltwater flavor of the surrounding air from outside; and, mixed with these, the fragrance of oilskins, the kind of heavy rain gear that John wore. The store contained every kind of equipment for fishermen: different weights of the cannonball leads and several sizes of the balloonlike red Scotchmen; a variety of hooks, lures, flashers, and high-test fishing line, with all the required small connecting parts (I recognized sinkers and tiny swivels) to put them together; floater jackets like mine for both men and women; orange and blue life jackets, life rings, and yellow and dark-green rain gear; rubber boots, T-shirts, flashlights, and all kinds of boat-rigging items with which I was vaguely familiar. There were ropes on big wooden spools on the floor—different sizes, different colors, including a lovely shade of bright blue; and big galvanized turnbuckles, like the ones John had, with a threaded bolt on each end that I had learned to twist to tighten the chains on the rail he had set up for me; and also pigs, their shapes slightly different from ours. These were like shoe boxes, made of both white and blue Styrofoam. One long wall of the store—it must have been thirty or more feet long—was partially covered with hoochies: a polychrome of little dangling bodies in beautiful bright, unexpected colors, like cerise and lime, blending with purples and reds; I couldn't think of a shade that wasn't there. The remaining wall space had an equally astonishing variety of spoons in plastic, shiny brass, or bright-painted colors.

I returned to my place against the outside wall by the laundry room, and pulled a *Times* I had brought along out of the laundry bag and read it until my turn came to wash. I put my clothes in the machine, and had resumed my reading by the time John arrived. There was a door on the other side of the laundry, from which he beckoned to me. "Imagine our good luck in finding the bathroom free," he said. "If we hurry, we can take baths and wash our hair before I unload my fish and the wash is done. I've brought your towel, soap, and clean clothes."

One look at the horrible, slimy cement on the bathroom floor and I said, seriously, "Why don't I rent a room in a hotel for an hour and take a bath? I am sure I can find some hotel that will let me."

"Nonsense," he said. By this time, he was tearing off sheets of paper towel from a roll he had been carrying under his arm, and spreading them on the filthy floor for me to step on, and two more sheets in the bottom of the dirty tub for me to stand on while I took a shower.

"Is it always this bad?" I said, and he grinned.

"Think positive," he said. "Think how clean you'll feel when you're through."

I started in a gingerly fashion to wash myself, and forgot all about the filth on the floor in the pleasure of washing my hair, after twelve days at sea. As soon as I was done and had put on my clean clothes, I ran back into the laundry room. My wash was spinning dry. When I had it safely in the dryer, I hurried around the winding floats to the *MoreKelp*, got the mirror and my portable hand hair dryer from the bottom of my duffel bag, and rushed back to the laundry room, where I had noticed a spare electric plug in the wall. When John came out of the bathroom, I was drying my hair before an audience of two fishermen waiting for machines. The clothes and I finished drying simultaneously.

While I was gone, John had moved the boat directly under one of the big cranes, in preparation for unloading. The big hatch cover was lying to one side on the deck, covered by a rising pile of green-and-yellow ice blankets that a young fellow was throwing out of the fish hatch, as he uncovered the fish. Another young man was scrubbing down the deck. They were a surprise to me: the unloading crew from the cannery, sent to help John, who had changed to rubber boots and the dark-green rubber overalls of his rain gear, under which he wore his usual Stanfield's underwear over a clean white T-shirt. The square aluminum bucket, suspended from a cable, was slowly descending on the winch from above, and, guided by John's hand, disappeared into the hold. When the boy had filled it full of the lovely fish I had watched John catch one by one, it slowly emerged through the hatch and was hauled up to the platform above.

After watching from the door of the pilothouse for a while, I became curious about what was going on above us. I timidly asked John, "Can I help?"

Without even glancing in my direction, he replied, "Stay out of the way, will you?"

I hardly thought I'd be missed, so I climbed the ramp again, which, thanks to the rising tide, was no longer formidable but at an easy angle.

The wooden flooring of the unloading dock was wet and slippery. I picked my way carefully along it, keeping close to the edge by the water after I was nearly run down by a forklift moving boxes of fish from the dock through the wide-open doors of the cannery buildings. I stopped for a moment to look down at our vessel below. The little square radar reflector standing on end on top of the port main pole was almost on a level with me and I could see straight down into the fishhold, which was already half empty. John, his gray hair standing on end, stopped to wipe his glasses, looked up and saw me, and smiled. From where I stood, he appeared to be helping his helpers, scrubbing down the blankets with a pail and deck brush, then rushing around to the bucket to tuck in a loose fish.

Three boats were unloading simultaneously below the three derrick winches. The young Japanese man whom I had seen from below was everywhere at once: like an acrobat, he flew from one job to another: from the hose, where he washed down buckets, to the winch, which he carefully maneuvered into the hatch of a boat just arrived; then a pause to consult with someone who had come out from the cannery; then back to the winch; then, waving his arms at an oncoming boat, to back away to allow a troller that had just finished unloading to leave; then, as the loudspeaker was calling out its name, waving another fishing boat under the vacated winch. He was a romantic figure, with his shining black hair, his slim physique in blue corduroy pants and high brown boots. John had told me that he was Ross Kondo, one of the four new owners of the cannery, recently purchased from its founder, a wealthy Vancouver businessman in his late sixties who had wanted to retire.

Ross nodded at me, and I walked over and said hello. "How do you keep it all straight?" I asked.

"I keep my eye on the boats as they come in," he said, "and we have a list. When their turn comes, we announce it three times over the loudspeaker. If they don't show up then, they go to the bottom of the list. There used to be fistfights at Prince Rupert and Namu about who went first, so keeping a list on a blackboard makes far more sense. If someone has to catch a plane—a lot of our gillnetters leave their boats here between openings and fly home—we will ask if you mind waiting, but it's never a problem. Everyone's very good about it."

Ross picked up a megaphone and shouted down to the boat that was moving into position under the winch next to John, and I moved along to watch a load of John's fish swing in the big bucket over to a large wooden table, waist-high, where a young man wearing a red cap opened a door in the side of the bucket, spilling the fish out onto the table. He jumped up beside the fish and began to sort them out with a long stick into big white tubs on wheels: one for the large mild-cure springs, and others for the medium and smaller ones, still others for coho, and for the pinks, or humpbacks, and for the occasional cod or halibut. When a salmon appeared to be borderline between mild-cure (twelve pounds and up) or medium (eight to twelve pounds) or small (under eight), it was popped swiftly onto the scale alongside to be weighed, and then thrown into the appropriate waiting tub.

It was hypnotic watching the fish-sorting, and I stood there for some time, until my attention was distracted by a new face that had appeared. A tall, rumpled, grinning teenager with glasses stood in front of me and said, "Hello!" Speechless, I just stared at the young man, until he said, "I'm Wilf Phillips. Don't you remember me?"

It took me a moment even then to remember that he was from Pender Harbour, and the eldest son of John's close friend Ray Phillips. I had felt so cut off from everyone but John in this fantastic experience I was having that I was astonished to see anyone now I had ever seen before. I finally said, "I certainly didn't expect to see you! What are you doing here?"

He cocked his head to one side, looking a little puzzled, and replied, "Fishing, of course. What did you think I'd be doing?"

"Is your father here, too?" I asked.

"Oh, Dad's here, but I've got my own boat this year, for the first time. The *Miss Derby*. I saw you up here and I just thought I'd come around and see how you were doing. How do you like fishing?"

"I love it!" I exclaimed, surprised at myself. I thought, I really do. "I'm watching John's fish get sorted and weighed."

The boy who had been sorting John's fish was pushing one of the white buggies containing the big mild-cure springs over to a scale set into the floor just inside the cannery doors, and we walked behind him. "The scale is set with the weight of the buggy, and that's subtracted from the total to get the weight of the fish," Wilf said. "My guess is that load will weigh around two hundred pounds." It weighed one hundred and ninety-five.

We had walked out of the cannery and over to get a good view of the unloading dock below. The bow of the *MoreKelp* was edging out into the crowded open space beyond, with a gillnetter poised to enter the spot it had vacated. We watched John move slowly and turn into the only space left to tie up, alongside three logs lashed together that I would have to cross to reach the boat. Oh, my God, I thought, one misstep and I've lost a leg between those rolling logs!

I mustn't think about it. "Is there anyone else here from the harbor?" I asked.

Wilf Phillips said, "See that green-and-white troller over by the oil dock? That's your neighbor Sonny Reid's *Instigator*." Again, a sensation of incredulity, seeing Sonny's familiar boat in this unfamiliar place. When I was in Garden Bay the previous winter, Sonny left his dock in Garden Bay every morning between five and six to fish for prawns on that green-and-white troller. John would say, "There goes Sonny," without raising his head from the pillow, and I could soon recognize the sound of the *Instigator*'s motor in the early-dawn silence myself.

Ross Kondo walked over. "John's been trying to get your attention," he said. "He wants to go now."

John was looking up, and I motioned that I was coming down. He nodded, and resumed talking to another man, who had his back to me. "Who is that with John?" I asked Wilf Phillips.

"My dad," Wilf said. "Everybody's here from home. They're mostly gillnetters." He went into the cannery, and I hurried down the ramp toward John.

As I came up, Ray Phillips was saying, "Wherever you fish, you like the place all to yourself, and when you get that, when nobody else is there, you think something's wrong."

John said, "I've never forgotten what one of my oldest friends, Buster Lansdowne, taught me thirty years ago. We had been fishing for a couple of weeks at Winter Harbour, off the west coast of Vancouver Island, and the fishing got worse and worse. All of us took off to fish in one direction, south, like sheep, except this one boat going the other way—Buster Lansdowne. I stopped and asked Buster where he was going, and he said, 'I happen to think there are fish up there, and none of you buggers know, because you haven't been there in the last ten days, so I'll make up my own mind, and it will only cost me fifty gallons of fuel.' Christ, I didn't see him again that season, and he got six thousand pounds. He said to me at the time, 'Never forget to look for yourself,' and all my fishing life that remark of Buster's has made a decision work for me: gamble a couple of days and decide for yourself. There are always conflicting rumors, and it's hard to go where others are coming from. Sometimes I've turned around and become a sheep, but I've always regretted it. The best example I know of what Buster said is when I took my son Dick to the Danger Shoals on September 4, in 1968, and in three horrible days all we got were two tiny fish. He was annoyed with me because he needed the money for university, and I wouldn't leave. The next morning, at the same place, we got forty-five springs, and forty or fifty cohos. The second day, we got forty springs and fifty-nine cohos, and the third day twenty-nine springs and fifty-three cohos—big cohos. The fishing gradually went downhill to nothing, but it was a damned good trip because of those three days."

John went off to telephone Mary Gunderson's husband, Nils, to come to the dock and pick up the springs he had for them. I asked Ray, "Do you fish springs, too?"

"I'm gillnetting, and right now that's mostly for sockeye," Ray replied. "If you're trolling sockeye, you might get the odd spring, but sockeye won't bite against the grain; you have to turn the way they go and almost dead slow, slower than slow, so you can't fish for anything else. If you're trolling with sockeye gear, you have to make up your mind that's what you're going to catch, and forgo the springs, which are the prize of the ocean. To concentrate on springs the way John does, you have to know a lot more, be a lot more patient, and obey a lot of rules some of us can't take for handling the gear. You have to wear gloves, and wash them to get the smell out."

Ray scratched his head. "Springs are the foxiest fish of all," he continued. "It just seems to take more expertise to catch them, and you have to have a boat that catches springs to be a spring fisherman. With John, it's a combination of him and the boat, because some boats just don't catch 'em. If you have a boat like the *MoreKelp* that catches springs, that's what you should go after."

Ray Phillips departed to unload his fish as John returned from phoning the Gundersons. He steered me to where the *MoreKelp* was tied up, walked across the three logs lashed together, and climbed on board. I stopped short, staring at those three logs. They looked slippery, and as if they would roll right over when I stepped on them; never mind that John had crossed without difficulty.

"Come on!" John said. "I'm in a hurry."

I stood there, frozen to the dock. When I rode horseback through woods in Ohio in my childhood, and my horse came up to a wooden bridge he didn't trust, he would stop, refusing to cross. I thought of that now. "I can't," I said. "I just can't. Those logs. I'm afraid they'll roll."

"Nonsense," John said. "You just saw me cross them without any trouble."

"I just can't, John," I said.

We stood like that a long time: John staring at me, and I, paralyzed with fright, looking at the logs. Finally, I said, "Why don't we find a dock I can walk to and you can pick me up there. Something solid to step *from*. Something that doesn't roll."

John leaped from the boat onto the logs and across to where I was standing. "Wait here," he said. "I'll be right back."

He returned, carrying a paper Lily cup in his hand. "Drink this," he said, handing it to me. "Drink the whole thing." I drank it down blindly, and then I did an amazing thing. With John holding my hand, I walked straight over the logs and onto our boat.

John instantly untied the *MoreKelp*, started the motor, and turned the bow away from the cannery. I came up beside him where he was steering, really puzzled. "What happened to me?" I asked timidly.

"Women and kids are like engines," he said. "When they start going hysterical, I've found that what they need is a good drink of cold water."

We ran for a short distance from the cannery to a red government dock, with many finger floats attached to it. John brought us in as close to the ramp leading from the road above as possible. He was tying up the boat when a truck drove up on the dock above us, and a short, stocky, elderly man with glasses got out and came down. John introduced me to Nils Gunderson, and handed him two of the four big white springs he had set aside in the fish box. John walked up the ramp with him, carrying the two larger ones himself, with a finger in the gills, and deposited them in his truck. "We'll be over to see you later," he shouted, as Gunderson drove away.

John untied the boat and jumped on board. He had left the motor running, so he backed up and turned the bow to head out. "We'll go over and tie up at Chris Sondrup's," he said. "If we stay here, we have to endure these horrible mercury-vapor lights all night."

It took only a few minutes to cross the bay to the other side. All I could see ahead on the shore was a small, neat red frame house, with two white-trimmed windows and a very odd roof. Four diamond-shaped shingled squares fastened together, slanting from the center peak to a point at each of the four house corners. The lower part of the house was hidden by an enormous rosebush covered with delicate pink flowers. "Look at those beautiful roses!" I exclaimed. "Whose place is that?"

"Why, Chris Sondrup's, of course," John said. "A little breath of beauty in this sea of ugliness." He was slowly nosing the bow toward a rickety L-shaped float and ramp made up of old boards and logs loosely put together, like ours at Garden Bay. We came alongside, and I saw an elderly wisp of a man with a fringe of white hair around his bald head, and large eyes in a round, elfin face, sitting in a little red rowboat, one hand on the dock, as if he were waiting for us. John put his head out the window, shouted, "Hello, Chris," and turned off the motor.

"I hear you, John. I saw your boat come in the harbor, and at the cannery," Chris said, speaking softly. His voice had a musical lilt, which I gradually identified as a Scandinavian accent. With this storybook house, he's right out of Hans Christian Andersen, I thought. "I wasn't sure when you'd get through unloading, but I always know that when you are in the harbor you'll be over."

Chris slowly got out of his skiff, and before he tied it up he glanced quickly sideways in my direction. I had been standing on the narrow side deck, gripping the pilothouse roof, and I dropped off onto the dock holding the bow rope until John came along to fasten it to the cleat. (It was a year before I trusted myself to tie the *MoreKelp* to a dock; I had a recurrent nightmare of standing by helplessly, watching it float away from a mooring because I hadn't fastened a knot properly, and I wasn't taking any chances.)

John went back on the boat, down in the hold, and came up with two coho and a spring, which he threw out on the dock. Chris shook his head. "Too much, John. These are good ones you can sell."

John laughed. "If you don't take them, my taxes will be too high, and the government will get them," he said. He introduced me, and Chris acknowledged my presence with a quick nod of the head, turning his face away, as if he wanted to hide. "I have to go back to the cannery, see Nils and Mary, and pay a little visit to the Spencers and the O'Connors," John said. "We'd like to spend the night here, if it's all right with you. Meanwhile, how about a cup of tea?"

"If you don't mind the way my place looks," Chris said, turning to follow John, who was carrying the fish. I had the remains of our red-snapper stew with me in a container for Chris, who stopped and, with a courtly bow, motioned me to go ahead of him.

We walked up the beach in single file, past an old fish boat on its side above the high-tide mark, and over some logs that John stepped across easily but that I had to scramble up and down. We passed several small buildings, a smokehouse, and a toolshed, mounted a flight of wooden stairs beside the rosebush, and entered the house through a side door.

"What a marvelous house!" I said. "I have never seen a roof designed like this. Who built it?" I was standing in the center of the one room, looking up at the intricate construction of the slanted ceiling sections into which rectangular skylights had been cut that flooded the room with sunny light.

"I did," Chris said, turning his full face toward me for the first time, and apparently forgetting his shyness in the pleasure of my appreciation of his architectural skill. "Who is there to do it for me?" He waved his hand to a bench across the room by a plank table with a blue, patterned linoleum tablecloth. Seated there, I had a full view of the room. The walls were paneled in plywood, into which odd-shaped cupboards had been built, and one nook with bookshelves was packed with well-worn volumes. His bunk was in the corner across from us, and a sink with one tap and a wood stove were along that same wall. There was a telephone but no electricity, and I saw only one oil lamp. Chris put a kettle of water on the oil stove, and made a half gesture toward picking up the clutter of clothes on his bunk, then shrugged and said, "I keep things in a mess. My good friends Judy and Einar Vagmar come over to visit, and Judy always wants to straighten things up—especially if she knows people are coming to see me. I say, 'No, no! If we want to see an animal, we should see it in its natural state.' "

After he had made tea and placed the pot before me to pour, Chris put crackers, a whole side of a smoked salmon, and a knife on the table. Smiling, he said, "My friend Tex Lyon brought me this fish. When I saw it, I said, 'I'll smoke it. Now maybe John will come.' "

John carefully cut off a piece and handed it to me. "You have never tasted smoked salmon like this," he said.

Chris laughed. "It's the good salmon that makes it that way. That was a fine coho." It was still warm from smoking and had a light, delicate flavor. "It's the wood I use," Chris said. "Old-growth spruce when I can get it; sometimes I use very dry alder, too."

While we were eating, Chris got up, nervously paced over to the bookshelf, took a book or two out, hesitantly came toward me, then went back and, while he was facing the shelves, said over his shoulder, "John has told me you are a writer. Perhaps you would be interested in these." He returned with an armful of books, which he piled up in front of me.

I picked up some books from the pile—one about Fidel Castro by a French author, one about Cuba in Spanish, and a third about Cuba's Bay of Pigs in English. "How many languages do you know?" I asked.

Chris made a dismissive gesture with his hand. "A little German, some French, English, and Spanish—enough to read but not to speak." He went to the bookshelf and returned with another armful of books: dictionaries in French, Spanish, German, and English. "I teach myself," he explained. "Languages interest me very much. Danish, of course, was my first language and all Danes have a familiarity with German. I learned English in school before I came here, and I still am not so good as I would like to be. I can read French with the help of this." He patted the French dictionary. "And lately I have been studying Spanish. That way, I keep myself so busy that I stay out of trouble."

"But you are also a fisherman," I said.

"Retired," he said. "Too old now. That's my last troller, the *Osprey*, we passed coming up from the beach. For many years, I had my own little sawmill here and used to build boats as a hobby."

"He is still building boats," John said. "He just finished that red skiff, and it's a beauty."

Chris rubbed his cheek nervously. "That's the end. I have no toolbox now," he said.

The edge of a long gray wooden box was visible under the bunk, and I said, "Isn't that a toolbox under your bed?" With a little laugh, Chris replied, "That's a bathtub. I built it." He reached down and dragged out a box about three feet long covered with dust, and turned it over.

It certainly was a bathtub, with smoothly planed, slanted sides and a small drainage hole at one end. "How do you use it?" I asked.

"I put it over by the sink and run the hose in and just let it drain across the floor out the door," he said. "It works very well, and washes the floor at the same time."

We all laughed. I looked from one to the other and asked, "How long have you known one another?"

Chris said, "We met in Vancouver, in Coal Harbour, in 1940, on my twenty-eight-foot boat, the *Cleo*—that's short for Cleopatra—named after a neighbor's cow when I had a farm. I didn't want to paint a long name on it. I lived winters on my boat in Vancouver, until 1948, when I bought this place in Port Hardy. I don't recall exactly when you started staying at my dock, John, but it was some time in the sixties, when Kennedy was President of the United States. I had gotten a few books from the library about Vietnam, and you were happy to see them, and took them to read. I had bought several books, to get the story of what was going on from different angles."

Canada is still a pioneer country—especially in the West—and when I meet a free spirit like Chris, with a touch of that pioneer quality, I am seized with a curiosity to know the whole story. "I suppose you came to Canada as a child," I ventured.

"No, no, I was twenty-two when I arrived in Canada, in 1924," Chris replied. "I was the youngest of twelve, and we had a small farm in Denmark. My eldest brother was a blacksmith. He went to the Dakotas in the United States before I was born, so I never saw him. I remember my father only as an invalid who had to be carried everywhere. My sisters took turns taking care of him after my mother died, when I was six. When my eldest sister, who was almost my mother, left for Canada in 1912, I went to live with an uncle on another small farm. Newspapers in Denmark were full of 'Come to Canada' ads and boat tickets were very cheap. Danish settlers had formed a farming community at Cape Scott, at the very northern tip of Vancouver Island, and my sisters and their families were among them. They were promised a railroad by the Canadian government to get their produce to markets, but it never materialized and the settlement eventually failed."

Chris paused. "Help yourself to more salmon, *please*. I can't eat all that myself," he said, with a gesture toward the plate, then continued, "I came on a little boat, six people to a cabin, that dropped us in Halifax, Nova Scotia, in March, when the snow was over everything. The train to Vancouver took five days. I remember the snow on the ground in the Rockies, but on the west side of the mountains the next morning everything was green, with skunk cabbage blooming, and the passengers went out and picked the pretty lilies. One sister and her husband were living on a homestead near Holberg, a small community that still exists between here and Cape Scott, and I found my way to them. For some years, I homesteaded in Holberg—built my own house and kept chickens, a cow, and a pig, and had a garden—and in the summers I fished from a skiff with a net by hand."

Chris went over to the stove, returning with fresh tea, and poured it into our mugs. He helped himself to more salmon and sat down. "The cannery I was fishing for in 1926 paid me twenty-five cents for a whole sockeye," he continued. "Once, my brother-in-law brought in a spring that weighed between sixty and seventy pounds. The cannery didn't want it, so he cut off a steak for himself and threw the rest away." Chris paused while he stirred sugar into his tea. "Another cannery I worked for used white spring for bait until someone got the idea of canning white spring with a label that read, 'Guaranteed *not* to turn red.' *That* went like hotcakes."

We laughed. "When I sold my farm, I built a bigger fishing boat and lived on it until I bought this place," Chris went on. "I guess the reason I am still here is that I have nothing else to do," he said, smiling. "I never planned anything in my life except what to eat for the next meal."

John looked at his watch and slowly rose. "We've got lots to do to get away again tomorrow morning. Can we take a turn through the garden on our way out?"

We went out through a door on the opposite side of the room from the entrance, down a small flight of wooden steps, and up a path that wound through colorful cultivated flower patches—roses, lilies, poppies, nasturtiums, and a few late tulips—encircled by wild floral growth. We stopped halfway up the hill at a shedlike structure with clear ribbed-plastic walls. "My greenhouse," Chris said, moving a section of plastic aside so we could enter. We walked around tubs of staked tomato plants, onions, and spears of corn just beginning to show. He raised his hand to the roof and slid a panel back to expose the sky. "This catches the rain for my plants," he said.

We stopped at his potato patch on the hill, where he pushed aside some weeds and lifted out half a dozen early baby potatoes. "As you see, these are grown in sawdust. From my mill," he said. "I don't have to wash them." At the top, we stood in a cleared area overlooking the bay, the cannery, and the town—pockets of low buildings set down on a sloped space with a wilderness of trees beyond.

"My picnic ground," Chris said. "I planted the clover, and the huckleberries and blueberries just grew." On the way down, he stopped again, dug a bulb from the dark earth, and handed it to me. "When you get home, put this autumn crocus on your mantel until it flowers. It has a pretty lavender flower. Afterwards, split the bulb into three and bury it in the ground. If you keep it longer in the house, it will be like a bird in a cage."

On the way to the boat, I lagged behind John and Chris, trying to analyze why this particular house and garden had such an aura. It was peaceful and untamed, like Chris. Running to catch up, I slipped on a mossy stone and fell hard on my back in the wet sand. When I stood up, my shirt and pants were soaked. Then I tripped over a wire holding the logs tied to the ramp, and fell on my face. The two accidents were swift, savage, silent, and bruising; I took them as a warning. I had escaped injury this time, but the next careless slip might be disastrous, so I had better watch myself. Walking stiffly along the ramp to the boat, rubbing my head, I had time to recover from a childish impulse to cry. I was grateful to get past Chris and John unnoticed while they were unfastening the lines on the boat, go down below to my nest in the bow, and change to dry clothes.

John crossed the bay again to the government dock, running the *MoreKelp* between floats, and backing it out again for half an hour, looking for a space to tie up. "What a bugger this is!" he exclaimed. "I hate being hemmed in by other boats." He finally gave up and went to the outermost float, where he threw a line over to the boat farthest out, a large troller, climbed on board, and fastened us to it. An ominous number of vessels were tied one to the other, between us and the dock. While John was putting a padlock on our pilothouse door, I started over our gunwale. I stood up on the edge, my right hand gripping one of our stays, and reached with the toe of my sneaker across the two-foot opening where our bumpers separated the boats, to the higher rail of the larger troller. I was balancing shakily, one foot on each boat, when John jumped into the big troller and gave me a hand over and down, from then on turning back to show me a footing, helping me up and across high rails. We climbed in and out of trollers and gillnetters of all sizes, and two big seiners, one after the other; down onto stern decks, up their sides to the next boat. Where a stern deck was so full of equipment that there was no place to stand, we crept along a narrow side deck amidships to the bow, then crawled over the roof of a tiny gillnetter, from which John pulled me up onto the deck of the next vessel. I lost track of where I had been or was going. When I slid down the side of the final boat onto the flat boards of the swaying dock, I turned and counted the vessels we had traversed.

"We have just crossed *eleven* boats!" I shouted at John, running to catch up with him.

"You're lucky it wasn't more," he said, grinning down at me as we hurried along.

We walked up the highway, stopping at the cannery office to get the slips for the fish John had sold. Scrambling down the rocks to the beach, we went along the water's edge to a path that led to the post office, where John mailed our letters and picked up mail for us. He opened a letter while we were walking and stopped in the middle of the road. "Good God!" he exclaimed. "Listen to this! It's a letter from Casey Lacina, the fellow who's house-sitting this summer while we're gone. It says, 'The first weekend after you left, I went down to start your car, and I'm sure glad I did it then and didn't leave it for another week. As it was, I could barely stand the odor in it and tracked it down to three lovely pieces of salmon on a paper in the back seat. After a week closed up in a warm car, they were *quite ripe*. However, I disposed of them, and no harm done.' " John folded the letter up and put it in his pocket. "Poor Casey!" he said. "I'll have to get an apology off to him before we leave. I remember now. I meant to drop those fish off at the store for Ivy Potts, the widow of our former postmaster, on the way out of the harbor, and I never even remembered to put them on the boat."

We resumed our walk, turned at the main thoroughfare, and came to a pink frame house. Instead of going up the steps to the front door, John led me around to the back yard, where Nils Gunderson and his wife, Mary, a small, wiry woman with graying black hair and snapping black eyes, were cutting up a large salmon that she had just finished smoking over a grate in the yard. She carefully arranged the pieces in shiny half-pound and quarter-pound gold-colored cans. Then Nils put the lids on, sealed them down with a small mechanical device, and carried the closed tins into the house, where Mary put them in pressure cookers on the stove, to finish the canning process. A covey of fish heads with glazed round eyes and gaping jaws were bobbing up and down in a large stewing pot on a back burner. "We're going to have our favorite dish—fish-head soup—for supper," Mary said.

"What do you do about the eyes?" I asked.

"Eat them! That's the best part of all," she said.

We sat down at the table in the Gundersons' big kitchen, in front of a plate of smoked white spring, served up with a loaf of Mary's fresh-baked bread and a pot of tea. John, with an ecstatic look on his face, held up a piece of salmon on his fork and said, "Before eating our first smoked white spring of the season, Mary, we should bow down and make spring-salmon noises such as I am making now, separate the chunks into flakes with our forks as I have just done, close our eyes while eating"—he closed his eyes—"and we can then travel that spring's life route down the Columbia, Fraser, or Skeena River to Alaska, where it will spend two or three years in the ocean, before it goes back five to eight hundred miles to spawn." He popped the salmon from the fork into his mouth, swallowed, opened his eyes, and continued, "Life begins with nature, not religion. The springs might be here to start it all over again, long, long after two-legged animals like us have rotted off museum shelves."

Our next stop was the supermarket down the street. John walked up to a brown-haired woman with glasses and lovely features presiding at the checkout counter. She smiled broadly as John came toward her, and he introduced me to Lorraine Spencer, Peter's wife. "We'll meet at your house and go out for dinner at May's place, O.K.?" John shouted happily, grabbing a shopping cart. He sped down and up aisles, dropping items into the basket. In less than ten minutes, while I was still looking around, he was at the checkout counter with a large order. While he was helping a delivery boy pack our bags into a truck, I hurriedly asked Lorraine, "Is there any place in Port Hardy to buy a cookbook?"

She reached over to the magazine and book rack and handed me a white paperback with a spool binding and five red roses portrayed on its cover. "This is put out by the Five Roses Flour Company and is the only cookbook I know of for sale in Port Hardy. It's a good one," she said.

I put a dollar and a half down on the counter, tucked the book into the side flap of my purse, and ran out to the truck, where John was impatiently waiting for me to climb in beside the driver. He dropped us and our grocery bags at the government dock, and we lugged them down a mile of ramps, and carried them across eleven boats, to the *MoreKelp*. John put the perishables into the white box and lowered it into the empty hold. "We'll be getting ice first thing in the morning," he said.

We started out again for town. "This walk will do your sea legs a world of good," John said. Beyond the cannery, we turned down a wooded road that ended at the big white house above the beach I had noticed when we were coming in the harbor. "We've just enough time to drop in on Dave and Lillian O'Connor before we go to Peter's."

On the way to the O'Connors', John pointed out his late fishing friend Terry O'Connor's house on the beach, an imitation-brick, asphalt structure. Terry's son, David, and Peter Spencer were two years apart in age, and had been friends, as well as neighbors, since boyhood. When we arrived, Lillian O'Connor, a blue-eyed young woman with auburn hair, was at the far end of her garden, weeding her vegetables. She had been teaching all day and couldn't join us for dinner, and David was out at one of his company's logging camps, she said over a quick cup of tea, sitting in a nook off the bright modern O'Connor kitchen in the luxurious home that David had built for them himself. We left Lillian's with John carrying strawberries and a big bag of rhubarb from her garden and climbed over the rock border at the shore edge of the O'Connor property onto the beach again. "Peter was the first white child born on this side of the bay, and this is where *he* lived as a child," John said as we passed the remains of a weather-beaten float house pulled up on the shore. Minutes later, on the bluff above, we arrived at the yellow-and-brown house that Peter and his father had built later. We were rescued from a confrontation with a large black dog on the porch, which was barking furiously at us, by the arrival in a pickup truck of Lorraine and Peter Spencer, to whom John introduced me. "I'm glad to meet you at last," Peter said, holding out his hand.

Anyone standing near John always seemed short to me, but Peter was probably close to six feet tall, a husky man in his forties with green eyes and long, dark sideburns traveling to the curve of a prominent jaw. He was wearing a locomotive engineer's striped cap at a sporty angle on his head; old worn pants; and an open green jacket over a navy wool shirt. He had a rapid, staccato way of talking, as if he were punching out each word. In addition to driving the oil truck in Port Hardy and being chief of the volunteer fire department, he was related to half the townspeople, both Indian and white. Nothing surprised him, and almost everything amused him. Even if I hadn't heard so much about him from John before we met, I would have liked him at once.

We went in the house, and when we sat down around the table in the bay window of the Spencers' dining room with a cup of coffee, I relaxed in the welcoming warmth of this extended family of John's. Lorraine told us she had seen our boat coming in from this very window as she was getting ready to go to work, and Peter remarked in an offhand way that he had finished renovating the *Diane S.*, the troller he had recently bought, so he guessed he'd go along with us tomorrow on our next trip; taking his sixteen-year-old daughter, Yvonne, for company on his inaugural voyage. The conversation drifted to our visit with Chris Sondrup and the smoked salmon we had just eaten there, which reminded Peter of "quite a yarn" about his uncle Tex Lyon, who had given Chris the salmon. I reached for my notebook and recorded it.

"This may sound implausible to you, but it really happened; I saw it myself," Peter began. "Tex is a man who is engrossed in the sea and sea life and he had this pet octopus. During the Second World War, around 1941 or 1942, he and a crew were unloading pipe fittings off a small boat for the construction company that was building our Port Hardy airport, when one of the cardboard boxes burst while they were slinging it over the side, aiming for the dock. All these individual black cast-iron pipe fittings fell into about thirty feet of water with a mud bottom; so there were black cast-iron fittings and black water. He and his crew tried for quite a while to get them, dragging with hooks, and then Tex thought about his octopus, and went out to the reef that's under the water in front of our house to find him." I must have looked doubtful, because Peter looked at me and added, "Even an octopus has a home. You can tell where its cave is because there are masses of crab shells and fish bones around it—a kind of garbage dump, you might say. All you have to do to catch an octopus, because he's not a fussy eater, is to stick something down, usually with a baited hook, and pull the hook up with the octopus on it. Which was what Tex did. Tex has an inclination to exaggerate and says now that he tied a rope around the waist of the octopus and brought it up, but it was a hook. If you ever tried to tie a rope around an octopus, you'd find out that you can't grab hold of a live one—it's a dozy, wet, slimy thing that slides—and then once you do get hold of it you can't get rid of it because it's got hold of you. And in so many places! That octopus was about six feet wide with eight legs, all of them three and a half feet long, and they have little suckers like suction cups on them. Tex took the octopus to the dock where the pipe fittings had spilled, put it in a boat, rowed out, and dropped it over the side. The octopus not only hooked on to the pipe fittings, but it brought up beer bottles, boom chains, tin cans, rocks, and gravel. You name it. Finally, after many many times of being dropped down and pulled up, the octopus brought up *all* the pipe fittings. That was very important to the construction company at the airport, because during wartime pipe fittings were so scarce. It would have taken at least a week and a half to get replacements from Vancouver. Of course, it wasn't all that easy. When you lower an octopus overboard, it will go wherever it can, so you have to figure out a way of enticing it to go to the bottom right below. I watched that octopus bring up those pipe fittings myself, and it was hard work." Later, I asked around, and found, to my amazement, that Peter's account was true.

It was still light out, with a warm glow from the waning sun, when we started off for dinner, with the four of us squeezed into the big crew cab of the truck. Our destination was the Pagoda Gardens, around the corner from the main highway, across from where Lorraine worked. A bare, noisy lunchroom, with a large clientele of fishermen and local Indian families, it was presided over by a tiny, energetic Chinese woman with very black hair named May Wong and her elderly husband, Dick, who was washing dishes when John took me in the back to meet him. It was John's favorite restaurant in Port Hardy, despite the fact that he was always watching his weight because of his heart. Whatever the entrees were supposed to be, they consisted almost entirely of noodles or rice, with tiny pebbles of protein, identified on close inspection as either meat or shrimp, buried in white hills of starch and surrounded by a sea of tan gravy. "They work all the time," people were always saying about the Wongs, and they were rumored to be millionaires. At least once a season, John presented May with a large salmon. I sometimes thought of the whole world—our whole world—as being divided into the people to whom John gave a salmon or did not give a salmon. "If they can afford to buy it themselves, they should," he once told me. May Wong was in a special category. He said that she deserved one, in recognition of the hard work she put in, all day and most of every night, at the Pagoda Gardens.

After supper, the Spencers drove us to the government dock, and we climbed over the eleven boats again, this time on tiptoe, since most of the owners were now on board asleep. I was several vessels behind John, and by the time I landed on our deck he had the Gardner running. With the engine purring softly, we proceeded across the shiny black water under a starry black sky to Chris's dock. All was still, and no light shone from the house, but when John went on the dock to tie up he brought back a bag of vegetables and a bunch of pink roses sitting in a bottle. Chris had left the vegetables and flowers for me.

John Daly in the stern of his troller, the *MoreKelp*, watching for signs of fish. He's holding on to a rope to steady himself, while the automatic pilot does the steering.

My very first view of the *MoreKelp*, on a marine ways at Garden Bay. I explored the pilothouse interior, while John was scraping paint off the hull underneath.

At Boom Bay, a storage place for old floats and fish camp buildings near Namu, where nobody else ever tied up, which is why John liked it. Our forty-one-foot troller was average size; does it seem small? That's John beside it, catching up on correspondence.

John's workshop, ramp and float, seen through the trees from above. He transformed a third of his working space inside the shop into a study for me.

Our house and front garden in Garden Bay.

Chris Sondrup took this picture in his house at Port Hardy. John had given Chris a salmon on our previous visit to smoke for us, and we had just finished feasting on it.

Meeting friends at sea. We were talking from our stern to neighbours from Garden Bay, Sam Lamont and Anne Clemence, who were at the stern of their beachcomber, the *Vulture*, when Anne snapped this picture.

Among the hundreds of sunset pictures John took, this was his favorite. He won a prize with it in a newspaper contest.

Minnie and Harry at home in Deserted Bay, at the far end of Jervis Inlet. In the winter, "Deserted" was the right name for it, but they loved living there.

Inside the *MoreKelp*'s pilothouse, facing the bow. I spent a lot of time at that steering wheel, or sitting in John's seat, left, looking ahead for logs or seaweed, while he fished.

Laundry day. Sea air and sun are better than an electric dryer, any time.

Interior view of the starboard side of the pilothouse, looking out to the stern. Note bottles of nuts, bolts and fishhooks, screwed to the ceiling. Cupboard below, left foreground, contained cooking supplies and liquor; open shelves above held a hodgepodge of pots, lures, clothespins and other necessities behind two-inch high edges. I prepared meals at the counter, and held on to its raised rim in rough weather. Out of sight, left, the sink and stove.

John steering with an old car wheel he installed behind the fishbox, while he waits for the fish to bite.

John coming from the pilothouse, heading for the stern to fish. That space on the starboard side was the only cleared walking area we had. The square box, left, is the hatch cover entrance to the fish hold, where we kept the iced fish.

My only attempt at cleaning fish. I took so long on two small coho that John never suggested I try again. The gurdies and cannonballs, or leads, are on my right.

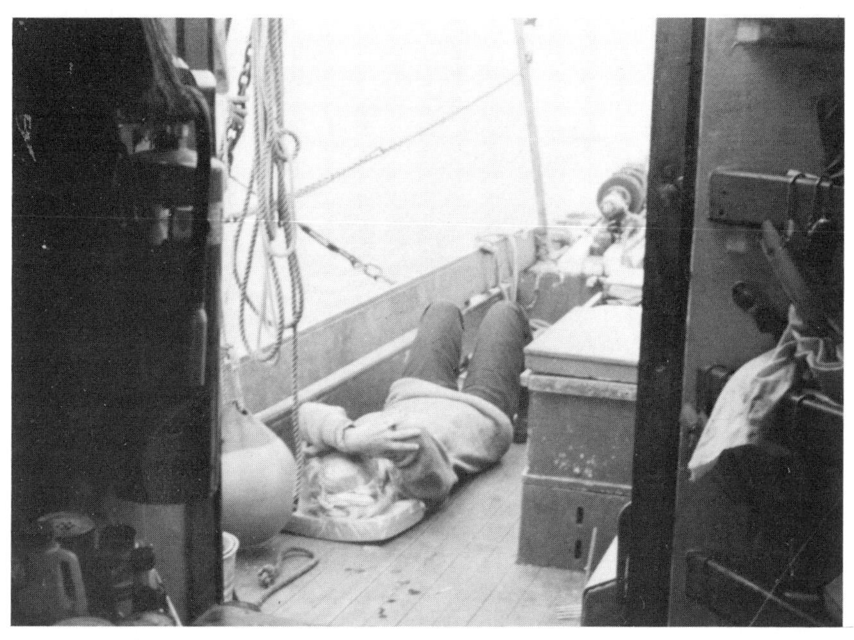

The concert hour: John's naptime. Favorite background music: Mozart.

John working the gurdies on a chilly day, with his cherished tartan wool hat pulled down over his ears.

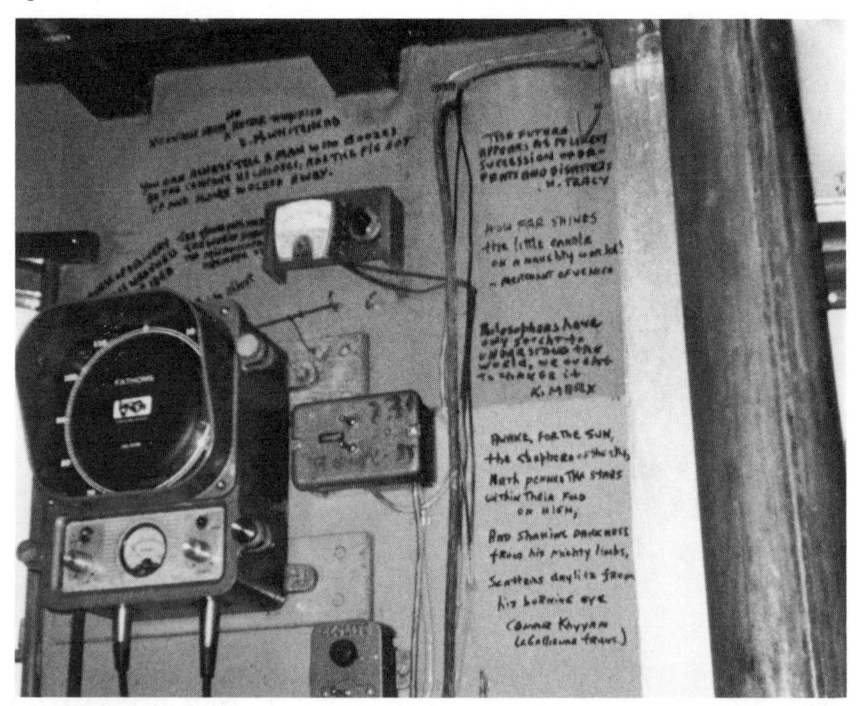

Inscriptions made by John with black laundry marker on white pilothouse walls. Excellent reading on dull days! Sample of starboard wall shows flasher-sounder, left, which indicates depth of water in fathoms, below vessel. Stovepipe is to the right.

Thirty-nine-inch spring being measured in the fishbox, prior to being cleaned—one of the big springs that "overflowed its bounds and overwhelmed my senses."

When we looked for a picture of the *MoreKelp* trolling, there were none; we were always on it. I took this of other trollers while we were all fishing slowly in a grand circle on the Steamer Grounds; so named because it was a favorite fishing spot for big steam trollers in the old days.

Picketing during the strike near Bella Bella, outside the Indian Fish Co-op, which was continuing to fish.

Seiners waiting to unload and sell their fish at Seafood Products fish processing plant at Port Hardy. Ice-making machine is at the right.

At the hot spring at Nascall Bay, standing in the doorway of the cabin, which contained a white porcelain bathtub, sunk in the floor, into which steaming water ran continuously.

The one-room red frame cottage Chris built for himself across from the cannery at Port Hardy. Described by John as "a little breath of beauty. . . in a sea of ugliness." Part of the lovely garden can be seen; the rest went up the hill.

Visiting Chris in his house: we have just made a trip through his garden, where he dug up autumn crocus bulbs for me to take home, and cut flowers to put in a jar on our boat. In early fall, not only my garden but those of my friends are filled with his lavender crocus flowers, to remind us of Chris.

Eucott Bay: that's Frenchy standing with me beside his motorboat.

John in his rarely-worn formal jacket and tie, on the only visit he ever made to my family in Cleveland, Ohio. Taken in the garden of my sister, Jane Fallon.

John, all smiles. Why not? He had just caught a very large spring salmon.

X

The alarm went off at five the next morning, and fifteen minutes later, while I was brushing my teeth, John slid the *MoreKelp* under the massive ice machine at the cannery, to be first in line. A pile of white boxlike structures towered above us. At six o'clock, a roaring sound erupted from the top of the pile as the ice tumbled down through a chain of what looked like interlinked pails into our hatch. When the hold was full to the brim with four tons of ice, with our white box of groceries perched on top, and our milk, eggs, meat, and cream buried along the edges, John laid down ice blankets as a cover, put the hatch lid on, and took the boat to the oil dock. He opened a pipe cap in the stern deck and inserted a hose from the water tap on the dock, letting the fresh water run into our tanks until it overflowed and ran out through the scuppers; twisted the caps off two pipes in the deck in back of the pilothouse and ran diesel oil through a heavier hose into our two fuel tanks, on either side of the engine; and filled the small tank that sat on our pilothouse roof with stove oil. I followed him into the store while he bought several packages of purple and fuschia hoochies, some diamond-shaped lures that struck his fancy, and settled his account. I wandered around the store, and my eye lit on one lone boat ladder leaning against the wall—the kind I had seen on the yachts that came into our harbor.

My heart was beating so noisily that I was sure John could hear it when I quietly put the ladder on the counter and said, "How much?"

"Twenty dollars," the clerk replied. I quickly paid and followed John out of the store, clutching my prize.

As I got on board, I slipped the curved grip of the ladder over the gunwale to make sure it fit. Perfect! We left immediately, without exchanging a word, but when I stood beside John while he was running out into the bay, he reached over and gave my hand a good squeeze.

"If you look out to the left as we pass the O'Connors', you might see Lillian waving a dish towel," John said. "She usually does, but it's a bit early." Sure enough, gazing up through the garden, I saw something white flash in the early-morning sun, and when I peered through the glasses, I could even see Lillian at the kitchen window. We both leaned as far out as we could through our wheelhouse windows and waved back.

We settled down then in our accustomed places, John steering from the big wooden bench and I sitting on the small folding seat. I could turn from there and guard the single open pot of coffee, so it wouldn't boil, and make toast on the wire rack over the hot spot on the stove, which was right behind me.

We came out of Hardy Bay into Queen Charlotte Strait, and John got out the charts and showed me where we were going: to the north end of Vancouver Island, past the weather station at Bull Harbour, across a blue area named Nahwitti Bar. Eventually, we would anchor at a place called Cape Sutil. Somewhere along the way, Peter Spencer would catch up with us in his boat, the *Diane S.* John pulled out the charts to show me our route. "I worry about you," he said. "You never seem to know where you are. You should be looking at these charts and know our position at all times, so that in an emergency you can tell people where to find you."

I worried about my ignorance, too; I was so tense about it that it was hard for me to concentrate when I looked at the charts. I took out my notebook and printed a list of all the radio call numbers I would need, and Scotch-taped it to the side of the big radio, next to the hook that held the microphone. I had a sudden thought. "What would I do if you went overboard?" I asked. "How would I pick you up?"

"I have never fallen overboard," John replied. "If I do, you must bring the boat around to wherever I am—slowly, so as not to turn it over. When you see you are close enough to me, move to the stern, throw me a rope, being careful *first* to tie it to the mast, and put that ladder of yours out on the side nearest me."

I had a vision—a waking nightmare—of John overboard and me out in the stern throwing him a rope. It didn't quite reach him, and I passed him and had to turn the boat around and try again. It was such a vivid picture that I felt faint. I could *see* his gray head and face, his glasses and white beard, bobbing above the water.

"You should also call 'Man overboard' on the Mickey Mouse, Channel 6," John continued, putting his hand briefly on that small radio overhead at the bow window. "Especially if all else fails." He had read my mind.

"I think it's time I tried the phone, so I get used to it," I said nervously. "Who should I call?"

"Why don't you try Sam Lamont?" John replied. "He and Anne are probably beachcombing, and Sam has his radio on all the time in case there's a log spill somewhere. Try them on the VHF."

"A great idea!" I said.

Sam Lamont and Anne Clemence had built a house about a quarter of a mile from John, and were close friends as well as neighbors. Anne was English, and until she met Sam had been a nurse at the peninsula hospital when it was situated in Garden Bay, long before my arrival. Sam's profession was beachcombing, which required a license and great skill to be successful, which he was. All along the B.C. coast, especially after storms, logs spilled off barges or logs that had come loose from the huge bundles constantly being hauled by tugs could be seen floating in the water—a menace to navigation. Sam and Anne were log salvagers, and, like firemen, were on call around the clock to go after the strays, which they picked up and returned to the insurance companies, getting paid for them. They worked from a tug constructed with very low sides that Sam appropriately had named *The Vulture*, which Anne could handle now as expertly as Sam. Anne was neat, small, quiet-spoken, and efficient, full of laughter and charm. Sam and his boat were well matched; *The Vulture* and Sam were short, wide, powerful, and efficient. Sam masked his intelligence, wit, and generosity with a rough manner and an impatience that could be withering, and I was a little scared of him. He and John liked and respected each other a lot but could not be in a room together more than ten minutes without having a shouting match.

I went over to the VHF above the bunk and picked up the microphone, hanging by a cord on a hook at one end, which I had seen John use. My mind went blank. "What do I do now?" I asked.

"We've been over this once before. Start by turning the left-hand knob on, to the right, and up for volume," he said patiently, and I saw the shadow of a smile when I reached over and grabbed my notebook again, carefully writing that down before I turned the knob. "Then adjust the next knob, marked Squelch, to get rid of static. Select your channel on that big knob to the right with numbers on it. As I told you before, you call out the distress signal, 'Mayday,' on 16, and give your location and this boat's name. The Coast Guard will answer and tell you to turn right away to their other channel, 26, so you are off the emergency line in case someone else needs it. Channel 21 is the continuous weather report from Bull Harbour; you just listen to that one. If you want to make an ordinary telephone call—say, to Garden Bay, or even to New York—you look B.C. Tel's channels up in the book. I think there are four, but Holberg is your nearest land station from here. We'll try Channel 13 to talk to Sam. Anyone can use that. Remember to press the little black button on top to talk, and take your finger off it to hear what the other person is saying."

When I had seen John call out, it had seemed easy. I held the receiver in my hand up to my mouth and, speaking carefully into the hole in the center, with my finger pressed on the button on top, said, "*Vulture, Vulture,* do you hear me?"

"Speak louder than that if you want to be heard. You're almost whispering," John said. "Say 'Over' when you're done talking, so the other person can reply. And be sure to take your hand off the button or you won't hear that reply. The chances are Sam *won't* reply, because God knows where those two are now, but this is a good way for you to practice."

The thought that I was just practicing and nobody would answer was a great relief. I was gripping the microphone so tightly that it was an effort to lift my finger from the black button. There was a crackling sound from the radio, but no reply, so I tried again, louder: "*Vulture, Vulture,* are you there?"

"*Vulture* here. Who wants me? Over." It was unmistakably Sam's voice, loud and clear, shouting above the familiar background racket of his 'Jimmy' motor, that noisy General Motors marine engine.

I was astounded to hear Sam's familiar voice coming through the built-in speaker in the VHF. I stood there holding the microphone, in shock. I don't think I ever expected anyone, let alone Sam, to answer—not the first time I ever tried a radiophone on a boat. "It's me, Sam," I said, but while I was still talking, Sam said impatiently—I could almost *see* his impatience—"*Vulture. Vulture.* I don't read you. Repeat, please. Over."

"He can't hear you. Press the button," John said. "Don't just stand there!"

I pressed the button. "*MoreKelp* here. Hello, Sam," I said faintly. "It's me. Edith."

Silence. I looked at John, and he groaned. "He's probably answering, but you've got your hand on the button, so you can't hear him. Take your finger off." Instead, I dropped the mike. It swung back and forth above the floor on the wire attached to the radio, and I could hear Sam shouting. All I caught was the word "Over."

I picked up the receiver, and put my finger on the button and said, "Sam? I forgot to take my finger off the button, so I missed what you were saying. This is the first time I've used one of these things. I thought I'd practice on you. I hope you don't mind. I never thought you'd answer." And then, hesitantly, "Over."

I took my finger off in time to hear Sam say, "Well, what did you think would happen if you called *The Vulture*? Who'd you think would answer?" Then he laughed and said, "You're doing all right for a first try. Everything O.K.? Where are you? Over."

I was trying to remember whether my finger was on or off the button, and which it should be so I could answer him, when I heard his voice again. "Press the button. You have to press the button to talk. Over."

I put my finger back on the button. "We're fine, Sam. We've just left Port Hardy. Over."

John pointed to the chart, and I looked where his finger was. "We're entering Goletas Channel. How's Anne and where are you? Over."

"Anne's fine and we're in Johnstone Strait. Congratulations on your first radiophone message. Everyone has to have a first time. And good luck fishing. We'll be off now. Roger." I heard the noise of the motor above his voice, and, in the empty silence that followed, I knew they were gone.

I put the receiver back on the hook and sat down heavily on the bunk. I was exhausted. "I'm glad I did that, but I never expected it to be so hard," I said. "I don't suppose it's ever as bad again as the first time. That was exciting, hearing Sam's voice." Reaching out into space and bringing Sam Lamont's voice into our pilothouse seemed like a miracle. I took a look at the chart to see where Sam was when we were talking. He was on a direct line with us but many miles farther south—how many depending on where *The Vulture* was in Johnstone Strait when we talked.

"I hope nobody else was listening," I said.

"Not more than half the population on this coast," John said. "Most people have trouble coordinating the button for talking and listening at first. You did all right. Considering."

We were in Goletas Channel now—a broad waterway with high

rock sides and a rugged shoreline. The bow of the boat cutting through the sea always threw back foamy spray, but this time there were added cascades of water roiling up just ahead. John stood abruptly, leaned way out the window, pulled his head back in, and shouted, "Porpoise! Get out to the bow and you can see them!"

I ran out the door and stepped carefully down the swaying narrow forward deck, gripping the roof with my hands. Surely they would vanish before I got there! I dropped to my hands and knees and crawled over the chains, ropes, and hatch cover in my way. At the bow, I straightened up and hung over the gunwale, straining my eyes to see the porpoise below me. A rush of water, and then I saw them, black-and-white streaks, dancing around the boat. There were a pair of them, moving in unison in a sweeping motion with incredible speed. They dove under the boat, their bodies slim and shining; came up fast on the other side. I had to look sharp to see them, splashing through the dark water, making the whitecapped waves sparkle around us. When they permitted a glimpse of their pointed snouts and gleaming slit eyes, their faces and rollicking forms, they radiated joy. Perhaps it was the curve of their broad mouths that seemed to perpetually smile, or the way they crisscrossed in front of the boat with such abandon. They made several more passes, and then disappeared. When I had given up all hope of seeing them again, they returned, separating to play a private game under and around us again, crossing from side to side, rising for a second or two to the surface. Then they were gone for good.

When I rejoined John in the pilothouse, he said, "They are often in Goletas Channel, I don't know why. It's probably got something to do with the feed here. They don't like to follow us when we're trolling; it's not fast enough or exciting enough for them. We see so few porpoises compared to the way it used to be! All fishermen are very sad when they catch one; I once got a half hitch around the tail of a porpoise with my line—that time, it was a dolphin, actually—and it drowned before I could get it loose. I shipped it off to Vancouver Island to the biological station at Nanaimo for scientific dissection, but I felt awful about it."

At the other end of the channel, John headed for open water to a place known locally as the Yankee Spot. He showed me an area on the chart—a 50-40-30-10-fathom contour in the shape of a thumb and finger. "Right around this thumb, that's the Yankee Spot," he said. "In the old days, it was a favorite place for American fishermen to congregate before the international two-hundred-mile limit for coastal boundaries was imposed. If we kept on going west for another twenty miles, we'd be on a very productive halibut spot we fishermen call the Steamer Grounds, because the big steam trollers used to fish there. The halibut steamers carried dorries nested on their decks that they dropped over the side with two men in them. Ray Phillips told me that a halibut boat called the *Celestial Empire* once landed one hundred and ninety thousand pounds of halibut there in less than twenty-four hours."

After a moment, he continued: "At the Yankee Spot, there are deep holes in its shallow waters, and usually you get an upwelling of feed in a place like that." He was looking through his glasses as we approached an area that was indistinguishable from the rest of the open sea to me. "Nobody at the Yankee Spot today! That means no fish. The last time I came, I counted seventeen boats." He had his lines in the water, and when he pulled them in with two huge springs on the hooks we both laughed. We were still laughing two hours later when we counted fourteen fish—the rest all good-sized, too—in the fish box.

While John was cleaning the fish, the sun disappeared, and it was suddenly cold. We were enveloped in a thick gray mist, traveling in the center of a gray igloo, confined by softly rounded walls through which damp air seeped to chill us to the bone. I put on my Stanfield's and windbreaker over my heavy sweater, and brought John his mackinaw and a red woolly hat with a white pompom which, pulled down over his ears, was the only bright spot in the grayness that engulfed us. The sea rolled past in slow, undulating waves, a heavy, heaving blanket of water that parted in ripples, with no whitecaps as we moved through it. Before the fog came, John had several times gotten out the glasses to look for Peter's boat, but now visibility was zero in all directions, and if there were any other human beings around, we could not see them. In this dangerous, hemmed-in world, we had only the little black dipchicks for company; they swam right up to the bow of the boat before diving, their pointed tail feathers a small triangle above the water.

John sent me forward to steer while he fished. "Keep to a north-northwest course and don't take your eyes away from looking ahead for a minute" were his orders. But in this cloistered, haunting space in which we were almost imperceptibly traveling, where all I could see ahead was opaque gray nothing, that was hard to do. John came in briefly to open the window because he said even the glass could obstruct my view of an oncoming vessel; and then the fog curled in like smoke. I had taken my Stanfield's off, but I began to shiver and had to put it back on. I continued to stare ahead into the gray mist, the silken curtain that enclosed us, until my eyes ached.

Boredom gave way to impatience, and impatience to a surreptitious looking-about on the compass shelf for interesting items; then to a ravenous hunger, which was hardly satisfied by chewing on some Britl-Tak. I said to myself, "Now I am a lighthouse." I swiveled my head left-center-right and right-center-left again, to cover any approach of another boat. The novelty of being a lighthouse wore off, and my eyelids drooped. I began reading the sayings on the walls, which I often did, almost unconsciously:

LAND IS A GOOD INVESTMENT
COZ THEY DON'T MAKE IT ANY MORE.
W. ROGERS

I had seen Will Rogers perform in vaudeville in my youth; I remembered his Stetson hat and the way he twirled a rope while he talked.

I speculated for the hundredth time about the missing words in the bottom line of a maxim partially obscured by the voltage-regulator box that had been more recently nailed over it. It read:

THE AIRPLANE HAS MADE THE WORLD SMALLER
THE MICROSCOPE HAS MADE IT

My head sank, and I jerked it up. I was fighting a terrible desire to take a nap. At one point, I thought I did drop off, and this was alarming enough to keep me bolt upright for another twenty minutes, until the whole cycle of ennui set in once more. Time passed. Hours of time. I was struggling again against sleep.

Objects, any objects that appeared within the confines of our gray prison, floating past the bow of the boat on that glassy gray surface, were a welcome diversion and took on special significance. A curved piece of light-brown wood an inch wide and about a foot long with a hole in the center of the slightly broader tip end moved in and out of my line of vision, to become an object of exquisite beauty and wonderment. It would drift toward us, then away, and then appear again at a slightly different location. I was fascinated. Who had made it? Where did it come from? Who had owned it, and what was it for? Had it been lost or thrown overboard or washed out from a beach? It disappeared for good, to what I thought of as "the other side" of that misty curtain.

I welcomed the arrival of an upright white plastic egg carton and an overturned half-pint plastic dish with a small blue-flower pattern around its sides—the kind used for margarine or sour cream—that rode in on the lazy ripples. No matter that they were someone else's garbage, not biodegradable; they were something to look at. I had never before appreciated the oblong shape and curved sections of an egg crate or the bright whiteness of Styrofoam; and the cheerful blue-flower decoration on the container reminded me that I must change the water in the bottle containing the roses from Chris. "Change rose water," I wrote down hastily on my pad, since I didn't dare move. I speculated on who had touched those castaways last. Someone from a boat like ours? From a sailboat? A sportfisherman?

A white gull with a black tail swam by, which I identified after a hurried glance in the bird book as a Bonaparte gull, followed by its winsome gray-and-white offspring. I broke up a round of Britl-Tak and threw it out the window. The mother made a pass at the fragments, and when the baby gull pushed her away she sat by and watched while it tried to eat. It was having an awful time, even though the pieces were small. It picked and picked at the morsels, making mewing noises like those of a human baby, while the mother fluttered impatiently alongside, occasionally devouring a piece in a single gulp—probably to show how it was done. I could follow the journey of larger pieces, swallowed whole, down her slender gull neck; they were visible bumps, first on one side, then on the other, on their way down. John came in for tea, and when I remarked on the eerie sound of the baby's cries in the mist he said, "God, they do that for three months, and they can drive you crazy with it if you are around them very much."

He left again, and I continued my vigil. A new fluttering noise outside made me glance around. A little brown bird had joined us and was walking uncertainly around the hatch cover. John, coming in, said, "We have a visitor. It's a poor little bird that lost its way in the fog. Sometimes several of those little land birds descend on the boat at one time, too wet to fly. I put them in the sink with the cutting board across the top and let them dry out, and when I hear them rustling as we get close to land again, I hold them above the pilothouse and let them go." The little bird made a few experimental trips into the rigging, moving always a little higher, and finally our brief caller flew away.

Perhaps I dozed; I could have sworn that I didn't. One minute, there was nothing ahead, and the next I was looking straight at a huge boat, a troller that appeared with terrifying silence a few yards from us on our port side. It looked gigantic. I didn't need any glasses to read its name: the *El Don*. Another troller, equally huge and menacing, came through the misty veil right behind it. They were so close! I thought, I can reach out and touch them.

I steered starboard in slow motion, not knowing what other vessels might be in that direction that I couldn't see, and shouted "John!" as loudly as I could, not turning my head. He had seen the boats already and was on his way in.

"God, I wish you had called me sooner," he exclaimed as he rushed into the pilothouse, grabbed the wheel, and steered carefully to the right. The two vessels vanished as silently and mysteriously as they had come, their stern ends receding through the fog that closed in around them. I remarked with relief that the *El Don* had radar; I had seen the small white receptacle on the roof with its antenna revolving. The two boats were traveling so close together that the *El Don* appeared to be leading the second boat, whose radar was not working; the antenna was motionless.

"That first boat had its radar on, so we really didn't have to worry so much, because he could see us," I said. "Isn't that right?"

"Never depend on that, because anyone could fall asleep or the radar could be broken," John replied. "Only the big tugs or the ferries carry two sets."

"Are we near any rocks?" I asked nervously.

John made a clicking noise of annoyance. "Can't you remember your chart? We're out several miles, and no rock piles, unless we were to go way off course. I still have to listen for any unusual noise, like waves breaking; watch the sounder, watch where the wind is coming from and the state of the tide by the direction of the ripples of water and the way the wood chips and garbage move past our boat; and of course be on the alert for other boats. I hope I'm blessed with a sixth sense about danger, but a person should never count on that."

John went back to the stern to fish, and for a long time I sat wide awake, watching the compass, watching the view ahead and beside us. The ghostly trollers appearing out of nowhere had cured my sleepiness. Then, without warning, the compass needle went crazy, swinging from right to left, then revolving rapidly in a complete circle. I turned the wheel slowly to the left, then to the right; the needle continued to spin, ignoring my orders. I felt the hair on the back of my neck stand up. "Don't panic," I whispered, frantic; staring ahead, glancing with beating heart at that awful whirling needle as I slowly moved the steering wheel a little from side to side in what I prayed was a steady straight course ahead, despite the contrary evidence on the compass that would have us turning in rapid circles. "Stop! Stop! Stop that!" I whispered.

Now I was soaked with perspiration. That terrible needle was still spinning. I took a deep breath, preparing to shout for John. An instant later, the compass needle was placidly moving between north and northwest, as if nothing had happened. I sat down heavily on the seat, opening my hands one at a time, to relieve the cramps in my fingers from gripping the wheel. It was then that the fog lifted.

Alone! We were anything but alone! We were in a populated area, surrounded by a dozen boats. No wonder John had been nervous; no wonder I, vaguely comprehending, had such a sensation of danger. Had I imagined the compass going beserk? No. Had the compass in the stern behaved the same way? I would soon know. I expected a scolding from John, so to get it over with I said, "I had a little trouble steering."

"You steer all over the place," he said with a grin, and changed the subject.

I figured that he must have been pulling in lines and didn't see it. (It was several years before I found out accidentally, while reading "Notes of General Interest to Mariners" in the *British Columbia Pilot*, that we had probably moved across magnetic minerals in the bed of the sea or a wreck of some kind in shallow depths that deflected the compass.) The sun came out to create a blazing sunset as we turned toward shore. John went below to ice fish, and I steered for the half-hour journey to our night's anchorage. Our destination was straight ahead, behind the black rocks of a natural breakwater, and as the sun danced across those rocks we entered the small protected harbor of Cape Sutil. We were the first in, but a dozen boats, including two sailboats, came in after us. As each boat stopped at a distance from the shore, a man would come out on deck and drop the anchor over the bow; when he went inside, the silence that his running motor and clanking anchor chain had disturbed would resume. Vessels were scattered eventually across the whole bay, several of them tied together for the night.

While we were eating supper outside on deck, a troller passed by, anchoring behind us toward land. John looked through the binoculars and said, "That's the Mad Russian. We call him that, although I think he's from Poland." Soon the glorious sounds of a familiar opera that I guessed was Mussorgsky's *Boris Godunov* but couldn't positively identify in these unexpected surroundings floated across the water: a whole gorgeous opera.

"Isn't that lovely?" John continued. "The Mad Russian always plays classical music on his tape recorder. He came to British Columbia after World War II, and in ten years he made enough money to buy a boat, a gift shop, land, and a house. When he first came, I gave him some spoons I had left over. They hadn't worked for me, and he went out with them and had wonderful luck; he got lots of fish. He asked me what to get for the next year, and I showed him in the catalogue. Well, the next season he came over to see me, totally grateful, but I hadn't had good luck with the same spoon order at all. So I asked him to show me his spoons, and it turned out that he had made a mistake and ordered spoons a fraction of an inch smaller and a slightly different color." He sighed. "That's the way it goes. He caught everything in the water with his spoons, and I suppose he felt he owed me something. One night, I heard a thump outside. He had dumped a whole lot of things for repairing a boat on my deck. 'Move this all down below as fast as you can, Daly, and disappear it,' he said to me. I thought it the better part of wisdom not to ask where he got it."

We watched the fog come in again outside the breakwater, enclosing all of us in the harbor in a single gray space, bordered by the shadowy line of the encircling shore. John groaned. "God, I'm tired. Here at Cape Sutil there are some particularly good specimens of carvings on the rocks, called petroglyphs, that are thought to be the work of prehistoric Indians," he said. "Nobody really knows who did them, but there are quite a few along the B.C. coast, and most of us—trollers, especially, because we stay so long at one time on our boats and feel the need to stretch our legs—like to put a dinghy over the side and go ashore and look for them. I wish I had the energy to show them to you now. Maybe tomorrow, if we still have this bloody fog."

An hour later, as the mist receded, the *Diane S.* arrived and anchored not far away. Shortly afterwards, Yvonne's voice came in on the Mickey Mouse. "Would you two like to go in the skiff with me to look at the rock carvings?" she said.

John replied, "I'm too tired, and I have to make up gear," but he accepted for me. Shortly, Yvonne arrived in a tiny green plastic skiff. Holding on to a rope John threw over the side, she waited while I hurriedly pulled on rubber boots and placed my new ladder over the gunwale. Wearing John's sleeveless floater jacket as a life preserver, I climbed down my new steps into the dinghy, and Yvonne, a big girl with an open, cheerful manner, rowed us with sure strokes to a sandy beach fronted with rocks. We came through green water so clear I could see a garden of ferns, reedy stalks, and flowerlike vegetation, undulating beneath us when I looked down over the side of the skiff. She maneuvered us between two boulders to the shore and I slid off the boat into a foot of water, dragging the dinghy by its painter up on the beach, as I had seen John do. Yvonne got out and we pulled the skiff above the high-tide mark, and tied the painter around a large, well-established rock.

We walked across a low grassy place and a wooded area to a beach of white sand studded with handsome gray and black boulders, and, below them, tree trunks and branches in all shapes and sizes, bleached tan, gray, and white from the sun on their endless journey back and forth with the tide over the rocks and sand.

Yvonne stopped, unsure where to go. "I came here once, with my dad. He showed me the rock with the drawings, but I'm lost now," she said. She started off again, down through a tangle of fallen trees and stumps, walked back in the opposite direction, and forward again through more tangled brush to where we could see the sea roaring ahead and whitecapped waves rushing toward us at the edge of the shore, climbing wild onto the black rocks that jutted out everywhere. Nowhere else have I ever seen such a stark contrast of black rock and white foam. The scenery was so dazzling that we sat down to watch the sea while Yvonne, with a teenager's fascination for the occult, talked in a low, dramatic voice, perfectly suited to our environs, about witchcraft. "Too bad I didn't bring along a pack of tarot cards," Yvonne said. "It would be just the place to read them."

Suddenly she stood up and walked straight across the rocks to a huge stone resting on its edge against the beach. "Come see the petroglyphs," she said. Along the rock side facing the land were two weathered faces, one almost round, the other somewhat elongated, chiseled into the stone. Each drawing had two round eyes, an oval mouth, but no nose or neck, and vertical lines at the tops of their heads that could have represented hair.

The petroglyphs were childlike, primitive; their mystery heightened by the waning light. Yvonne's conversation made them even a bit threatening, and I shivered as she continued to talk about witchcraft, weaving a new narrative in which the petroglyphs were incorporated.

I sat down in a smooth, round indent, like a seat, on the other side of the stone. It was a tight fit. "Do you think the Indians used this for ceremonies?" I asked.

Yvonne's greeny-blue eyes gleamed. "Or sacrifices," she intoned.

We lingered as the sun went down, I crouched on the ceremonial seat, Yvonne leaning against rock, the surf pounding, the spray and splash of waves making a great show for us. It was dark when we started back through shadowy woods on a half-remembered shortcut Peter had shown Yvonne, stumbling into trees and among brambles and salmonberry bushes to reach the other side—the beach where our skiff was. When we tried to leave, we found we were stuck on a rock. Yvonne plunged into the water in her boots to push us out, and I simultaneously jumped from my seat into the surf to help. The skiff, suddenly light, moved off the rock, and Yvonne, still pushing, fell on her knees into the sea, up to her waist. She came up splashing and, to my relief, laughing. No apologies: we both climbed into the skiff, continuing to laugh on the short row back to the *MoreKelp*. John was leaning over the side, waiting for us. Yvonne threw him the painter and he grabbed it and tied us up. My ladder was still in place, and I climbed up easily and over the gunwale, followed by Yvonne, who took off her sopping jeans and wrapped herself in a dry towel.

John handed us each a cup of steaming cocoa, smiling broadly while we breathlessly narrated our adventure. "Peter and I have been shouting across at one another—we were about to come looking for you," he said. "Meanwhile, I've dropped the crab trap off the side of the boat with a very tasty cod head in it for bait. Let's hope we have crabs for lunch tomorrow."

XI

In the morning, John got up as soon as it was light, took one look out the window, and went back to bed. The fog had settled in again outside the harbor, and nobody was moving. I heard him turn on the radio for the eight o'clock weather report and CBC news. Then he went out to pull up the crab trap.

When not in use, the trap, a circular black plastic narrow-slatted cage, about two feet in diameter, with a flat bottom and a rounded domelike lid, traveled with us tied down on the roof of the pilothouse. It had been resting overnight on the seabed, about eight fathoms down, on a long, heavy rope, whose other end John had fastened to the rigging on the port side of the *MoreKelp*. He hauled the rope up, hand over hand, and the minute the trap bubbled up to the surface of the water we could see that it contained something impressive. He lifted the trap over the side and dropped it with a thud on the deck. I jumped back, amazed. One enormous greenish-brown crab, about a foot wide, filled the cage. The crab was furiously crashing about, and while I remained at a cowardly distance, John opened the trap door, reached in with his rubber-gloved hand, and pulled the crab out by the back legs. He had interrupted it at dinner; there was a ghoulish-looking unfinished half a cod's head still in the trap. John held the crab high up and away, while it flailed wildly in the air with its huge claws. He shouted, "Bring me the teakettle!"

I ran and got the kettle from the stove. He dropped the crab into a pail, and as I poured boiling water over it, the unhappy creature sank down in its scalding bath, already beginning to turn the scarlet color it would have when it was cooked. I followed John into the pilothouse, where he put the pail on the hot spot on the stove. "That one will take about forty minutes to cook," he said. "Dick and I once caught an entire pail full of crabs here. We were boiling them all day."

I thought: This one will be quite enough. I wrote later that morning in my journal: "I have no character. I hate taking any life, but I *love* seafood."

When the crab was cooked, I picked the meat from the claws with a hammer and screwdriver, while John cleaned out the body. We had almost a pound of crabmeat and we planned to save some for supper, but it tasted so good that we kept making more buttered toast and taking just a little more crabmeat, until we had eaten it all.

The sun struggled all morning to shine through the fog, and by noon it had finally burned it away. All the vessels except the sailboats left Cape Sutil and commenced fishing. We went back to the now crowded Yankee Spot, joining the circle of other trollers, with the *Diane S.* a distance of several boat lengths behind us.

I fixed a quick lunch, one of John's Britl-Tak, cheese, peanut butter, and cucumber concoctions, which I was becoming fond of myself, and took it out to him as soon as all his lines were set. John pulled his eyeshade down firmly on his forehead and leaned against the stern, his arms stretched out along the gunwale behind him. "This sunshine!" he exclaimed. "I feel as if we've not see the sun for days. I'm *drunk* with this hour of sun."

He opened a bottle of beer and offered me some, but I shook my head. "Beer makes me sleepy, and I'm going to start writing," I said. "I thought I'd try that typewriter you said you had on the boat. Unless you need me to watch for you. Do you?"

"Certainly not!" he said, beaming. "What do you think I've been waiting for? I was wondering when you would get started. Me fishing and you writing! A purrrrr-fect combination!" While he was talking, he was taking off his gloves and climbing out over the fish box. He strode into the pilothouse, bent over, and began unwinding a rope around a tan leather case that was tied flat against the wall underneath the bow seat. I had sat there innumerable times, with my feet pressed against something hard that I thought was part of the boat construction, without noticing the object he was now holding up for me to see. It was the typewriter.

The rope around the typewriter was fastened to two hooks on the wall at either end to hold it in place. The rope also kept the bottom and top of the case, which had lost its hinge, together. John lifted out a light-blue portable machine that was in surprisingly good condition, and, holding it in his arms, said, "Where do you want to sit? Outside or in here?" We both looked around. John hurriedly glanced out the window, gave the steering wheel a quick turn, and ducked low to look out the door to see how his lines were. He turned back to me still smiling and hugging the typewriter in one arm, keeping a hand on the wheel. "It would be lovely to have you outside with me," he said, "but I don't know exactly how we could set you up so everything was solid; or what we'd do when it rained."

"Or when the wind blew my papers overboard," I said. "I'd better work in here. I'll sit on your bunk and use the table for my typewriter. It's the only possible place."

I unhooked the table and pulled it down. John put the typewriter on the table and dashed out the door, shouting over his shoulder something about tangled pig lines. I ran to the door, but he was already back in the cockpit, had brought the boat around, and was bringing in a line. He gave me a smiling nod and I went back to survey my new work space. The typewriter sat very firmly on the plastic table mat that kept our dishes from sliding around. I went down to my bunk and collected my notes, typing paper, pencils, erasers, scissors, and Scotch tape, and put them in the bag in which I transported belongings from my quarters below up to the pilothouse in the early mornings. I sat down sideways on the bunk the way I did when we ate. It was impossible to type sideways. I turned and put both legs out straight in front of me on the bunk, and, with John's pillow behind me, tried to type; but I was too far away. I went down below and brought up my bunk pillow and put that behind me. Still too far away. I added my floater jacket, one of John's Stanfield's underwear tops, and a gray wool blanket, all folded double behind me. Better, but still not close enough. I looked around. The dishpan! I popped that on its edge between the two pillows. Just right. It gave me hard support and brought me up square with the table. Two bed pillows, one floater jacket, one Stanfield's, one blanket, and a dishpan: they were my seat backup for two years on the *MoreKelp*, until Sam Lamont built me a wooden seat that folded up flat, with a handle for carrying. (When I opened it, I pulled out two wings hinged to the seat back that went a measured distance to the wall of John's bunk, taking the place of the dishpan and pillow arrangement. My "chair," a wooden back and seat, was then flush with my typing table, and I sat with my legs straight out under it, as before, but with a lot more support behind me. The best thing about Sam's folding seat was that it was all in one piece. Before I had it, the crash of the dishpan on the floor was often my first intimation that we were heading into a storm or a groundswell. Leaning down to pick up the pan, I invariably brought down the pillows; and, inevitably, John arrived as I jumped up to retrieve the whole mess strewn in his path. Keeping the aisle clear was in the same rule book with hanging cups up in one direction only.)

The typewriter, which John had picked up years before at a sale, performed extremely well. Every morning between nine and ten, except during storms—when the boat pitched violently—or during fog, or when John asked me to steer, I brought up my writing paraphernalia and began work, which I often continued until the roaring sound of the throttled-up motor signaled the end of the trolling day and the beginning of the run into a harbor. Then I would hastily gather up my belongings and take them below to the foot of my bunk, since there was no place to store them in the pilothouse, and tie the typewriter into its case back under the bow seat. Everything had to be in its place, so I could be ready to steer while John iced fish on the way in. That routine became automatic.

On my first day with the typewriter, we fished at the Yankee Spot again all afternoon. While we were having a rest, John picked up the binoculars, looking for birds. He scanned the sea to the west, and handed the glasses to me. "See that island far out beyond the others?" he said. "Beyond the nearest point, Cape Scott, the westernmost tip of Vancouver Island, where I have done some of the toughest fishing in rough weather you wouldn't believe, are those stone outcroppings we call the Haystacks, because they look like haystacks, although they are the Scott Islands on the map. The biggest and closest one to us is Cox, and the one farthest out, forty-two miles from Cape Scott, is Triangle Island."

Even with the glasses, I could barely see a lonely speck on the horizon in the open sea. "In 1910, the government tried the impossible, and put a lighthouse and wireless station, six hundred and fifty feet above sea level, on the bare rock of Triangle Island," he continued. "I don't know how many men—four or five, at least—have lost their lives trying to get supplies out to it. The idea was to give light to boats coming from the Orient, but Triangle Island's got the worst weather imaginable—horrible riptides and hundred-and-twenty-mile-an-hour winds that once blew a storage shed and cow right off the island! Most of the time, you couldn't see the light anyway, for the fog and mist around it, but it could sure give weather reports. In 1920, the government gave up on having a manned lighthouse there, and the weather station is now at Bull Harbour. A couple of years ago, when I was fishing off Cape Scott, somebody got on the radio and said, 'Holy Christ, you wouldn't believe what has just come in here at Cox Island!' It seems that three Indians fishing halibut in a rather decrepit old boat came around Triangle Island and anchored the boat while they went ashore in a skiff to collect seagull eggs and other delicacies, and the wind came up and wrecked their boat. They had a fair amount of grub with them, fishing lines, and one tiny little skiff that would at most take two. They didn't want to leave anybody behind, so they built some kind of a craft from floating logs that would hold one man, and they all three set off for Cox Island, because they knew that trollers would be anchored there. Two men were in the skiff and one sat in this floating box they had built, with some cans they had picked up on the beach. He just sat in there, wet through and through, all the time, and bailed, while the other two rowed, towing him. It took them fifteen hours to reach Cox Island, and they went up to some troller, who just didn't believe they had rowed all that way and were alive. He took them to Bull Harbour, where they got a ride back to Port Hardy. Not only did they survive, but it was a remarkable piece of seamanship."

John went back to fishing and I to my typewriter. Late in the afternoon, I heard Peter Spencer's daughter, Yvonne, calling me on the Mickey Mouse. I disengaged myself from my pillows and dishpan, bringing them to the floor with a crash, and answered. "Say, I've got the best chicken recipe, with wine and mushroom soup, and I've got the chicken, the wine, and the soup, but I don't know how to use the oven on this boat," Yvonne said.

I explained about pulling up the lever at the back of the stove to get the oven working. Then I asked, "What's the recipe?"

"You brown the chicken first and—" But before Yvonne could go any further, a man's voice cut in and said, "That sounds good, girls, and I wish I could have some, but now will you get off the phone?"

"Sorry about that," Yvonne said, sounding subdued. "Goodbye for now. Over and out."

At slack tide, I took tea out to John and watched him pull in his lines. He caught two big springs and a dozen fair-sized cohos. He told Peter Spencer over the Mickey Mouse that we would anchor off Egg Island for the night, and was heading out the door to ice our fish while I steered, when we heard Yvonne calling me. "How many fish did you catch?" she asked. "We got lots of coho and six smilies. There are *lots* of fish here."

"Good God!" John exclaimed. He wheeled around and came back in. "Don't say a word over the phone," he said to me. "The whole fleet's listening! In an hour and a half, every boat in the area will be out here. There's a saying that it takes news ten minutes to travel from Prince Rupert to Victoria!" I must have looked puzzled, because he added, "That's by radio from one end of B.C. to the other on the fisherman's band."

"I can't hear you," Yvonne persisted. "How many smilies did you get? Did you hear me say we got six?"

"I heard you, Yvonne," I said weakly. "Yvonne, how do you feel today?"

"What's the matter? Is your phone O.K.? I asked how many smilies you got," she replied.

"I'm fine," I said.

John grabbed the receiver. "Is Peter there?" he asked.

"He just came in, John," Yvonne said. There was a pause. "I'm not supposed to talk any longer."

"Fine. Good," John replied. "See you at Egg Island. Over." He hung up.

We had to cross through the heavy traffic in the shipping channel in Queen Charlotte Sound to get to Egg Island, and Peter and Yvonne traveled close behind us in the *Diane S.*, which skimmed along like a graceful bird. John was always tense crossing over from the Vancouver Island side of the Sound to the islands off the mainland; he stood firmly behind the wheel, steering with both hands, shoulders hunched, head forward, eyes darting in all directions. I knew better than to attempt conversation en route, and sat silently on the stool beside him, keeping a lookout for kelp and logs while he concentrated on avoiding other vessels—especially tugs with long towlines.

When we were safely across, John sat down. "My God, the traffic!" he said. "In the last hour and a half, we have passed a tug with a log barge in tow; an empty tug—that is, with no tow; an Alaska crab boat that I meant to call to your attention; three packers; four trollers; another tug hauling a barge; and two gillnetters. If you weren't with me, I would probably be trying to wash clothes, cook a stew, and write to you while I was steering among them."

As we went by Egg Island, coming at it from the south, I saw the lighthouse—with its metal tower and flashing light—which, according to the chart, was two hundred and seventy-five feet above the water. I had the conventional romantic view of lighthouses and commented on the handsome sight it made for mariners. "It's had quite a history of horrible events," John commented. "In November of 1948, the entire station was blown right off the island by winds coming straight down Hecate Strait for three days and nights. The lighthouse keeper and his wife and child had to struggle out of their bedclothes and run for their lives minutes ahead of a bloody big tidal wave that washed the house away, and they survived for five days in the bush before they were rescued. That's just one of the Egg Island horror stories, before the station was rebuilt on higher ground, as you see it now."

We came in behind Egg Island, and as soon as both boats were anchored, in front of a small beach, Peter Spencer called on the Mickey Mouse to suggest a picnic. John sighed. "I'm just too tired," he said. "Why don't you and Yvonne come here for a feed of salmon?"

"Good," said Peter. "I'll barbecue a sockeye on the beach and bring it over."

My job was to prepare the rest of the dinner, so I went out to get supplies from the white box in the hold. I saw Peter and Yvonne start off in their dinghy for a small strip of a little sand and a lot of rock which passes in most places on the B.C. coast for a beach. They landed, tied the skiff's painter around a rock, and began gathering driftwood. In no time, Peter had a good fire going, and while Yvonne continued collecting wood, I watched through the glasses while Peter took out his knife and whittled at some sticks, which he told me later were cedar. Then he split the sockeye from the back to the belly cavity, laid it open with the two halves side by side, and placed them on a spit that he fashioned from the sticks. The salmon was about five pounds and maybe twenty inches long, and I continued to watch as he wove six small cross sticks across the body of the fish, like a lattice, three on each side. John reminded me that I was supposed to be getting supper, so I hurried in to peel potatoes and start them boiling with the mint that John always brought along from the garden. I opened two cans of baby peas, which I drained into a pan that I set on the stove, adding shredded lettuce, some of our good canned butter, and a small onion cut up for seasoning, and made a lettuce, tomato, and avocado salad, using one of the seven avocados John had bought in graduated stages of ripening when he shopped in Port Hardy. I had made John another lemon cake while we were fishing at the Yankee Spot, and that would be our dessert.

When I had a moment to go on deck again, Peter was standing on the beach, turning the salmon from time to time. Yvonne was sitting on the opposite side, throwing stones haphazardly into the water, and the two of them looked like a painting of the Great Canadian Outdoors: Peter, standing by the orange-crimson coals, in a red plaid shirt, with his striped engineer's cap on the back of his head; Yvonne, brown-haired, braced casually against a boulder, in a red jacket and bluejeans. Both were patches of color against a gray-and-black setting of rock, and above that, brown tree trunks topped with heavy dark green foliage.

I looked at my watch: the sockeye had been cooking for half an hour. I ran in to get cutlery and dishes so that we could eat on deck, and when I came out Peter and Yvonne were on their way to us in the skiff. Peter climbed on board holding the skewered sockeye high in front of him, and set it carefully down on the hatch top. Yvonne produced a loaf of Lorraine's bread, and our feast began.

Barbecued salmon has a special taste; it is the flavor of salmon, wood, smoke, flame, ashes, and whatever seasoning the cook secretly applies, with a dash of the atmosphere in which it has been prepared. When I made glowing remarks about its tastiness, Peter laughed. "Heat and smoke on fish makes it just right," he said. "I cooked it Indian-style, my ancestors' way. My great-grandfather was a Scot named Robert Hunt, who worked for the Hudson's Bay Company at Fort Rupert, near what is now Port Hardy. He married the daughter of Princess Shining Face Copper, a Tlingit from the Nass River near the Alaskan border, and *their* daughter, Jane Hunt, became my Granny Cadwallader. She met my grandfather, Harry Cadwallader, in Victoria when he came out to bring the coast Indians back home who had been performing at the Chicago World's Fair in 1893. My Cadwallader grandparents had nine children. Their daughter Annie married my father, Dewey Spencer, and that's where I come in—me and my two sisters, Rosie and Faith."

Peter and Yvonne stayed about an hour and a half. It was still light out when they rowed off in the skiff to the *Diane S.*, but everyone was tired from the long day. A small orange troller, the *Tide Rip*, came in after dark. Its owner was a young friend of John's, Dave Hardie, a schoolteacher from Vancouver Island, who had become so enamored of trolling that he was about to make a full-time career of it. He rowed over to say hello, and stayed to finish the remains of the salmon and cake.

As he was leaving, Dave said, "I'm going to tie a line to my rail, bait my halibut hook with a big blob of salmon eggs before I go to bed, and throw it over the side. Maybe I'll catch another one-hundred-and-seventy-one-pound halibut. I got one that way around here when I was a kid."

At 5 a.m., we were awakened by the crack of a rifle shot that rang through the harbor. We heard a yell, "I got one! I got a sixty-five-pound halibut!" We ran out in the gray dawn, and Dave was jumping up and down on the *Tide Rip*, waving his arms. "Whee-hee!" he yelled. "Whee-hee!" He bent down and, struggling, lifted an enormous halibut for us to see. The Spencers had run out on deck, too, and we all shouted our congratulations.

The Spencers and Dave departed for fishing areas a little farther north, and we went back to the Pearl Rocks to fish. For a while, John baited our halibut hook, a deadly-looking three-pronged affair, and dropped it over the side at night without any luck. The Spencers and Dave would be delivering their fish into Port Hardy, and we were going in to Namu, so we didn't see any of them again that trip.

XII

Our main reason for going to Namu, John told me, was that he thought the United Fishermen and Allied Workers Union, in which he had been an active member for many years, was going to call a strike against the fish-packing plants. John was eager to join his fellow union members at Namu, where he would also sell whatever fish we might catch before the strike began. The huge conglomerate, Garfield Weston, through B.C. Packers, which it owned, represented close to half the total fish production in British Columbia. A handful of other companies picked up the rest. Over the years, largely due to the short fishing season and the fragile nature of the product, a cat-and-mouse bargaining pattern between management and labor had been established. This led to frequent strikes—sometimes every other summer. Contention usually centered around wages negotiated in labor contracts with shore workers, and the floor price paid for roe herring fished before the salmon season opened and for salmon, where the price varied according to the species.

At the time I was fishing with John, I knew only what I absorbed as I went along about the fishing industry in British Columbia. I was too busy making my own adjustments, being a fishwife, which John liked to call me, to investigate anything beyond my personal experience. The rest, the general situation for fishermen, was a blur; I could sort the facts out only later when I had more time. When I did, it surprised me to find that John was one of relatively few trollers who belonged to the union. As he said to me often before I realized the significance of the remark, "We trollers are a curious independent lot. Most of us are trollers because we *like* to be alone, and don't want to have someone else telling us what to do, even though it means working a fourteen-hour day."

I was startled by other statistics I uncovered subsequently about the commercial fishing industry. Numbers have remained fairly constant because of tight controls by the federal government, and although anyone now fishing in B.C. waters over the age of sixteen must have a personal sport or commercial license, the significant commercial license is issued on the boat. The number of boat licenses that are granted is decreasing, not because of a waning demand for them but because constant improvement in the efficiency of harvesting methods has become the greatest threat to the very survival of the resource. The government periodically has stepped in to eliminate licenses and buy back fish boats to remove them from the system, and they are then sold off at a nominal price as pleasure boats. Generally speaking, around 350,000 people sportfish in the salt water for salmon in B.C. in any year, compared to at least seven thousand active commercial salmon fishermen. Both groups use sophisticated electronics for locating fish.

When we were fishing, of the approximately seven thousand commercial fishermen, about fifteen hundred were trollers like John, of whom two hundred were in the union; about twenty-five hundred were gillnetters, some with combination boats, a thousand of them union men; and about three hundred of the five hundred and fifty seine boats had union crews, with an average of five deckhands per vessel, for a total of around sixteen hundred union members. The biggest single group in the fishing business—forty-five hundred shore workers—accounted for thirty-five hundred of the union's membership; and there were three to four hundred union tendermen working on the packers. The Native Brotherhood, about twelve hundred Indian fishermen, sometimes supported the UFAWU by tying up during a strike, and some of their members paid dues in both places. Some nonunion trollers stopped fishing during strikes in support of the UFAWU, and others tied up because the plants where they usually sold their fish were closed, without shore workers to process the fish. A few fishermen ran to the United States, to Washington, to sell their fish, and the Prince Rupert Fishermen's Cooperative, which accounted for between seven and eight percent of fish production in B.C., kept fishing. Its plant stayed open; its shore workers had their own union.

While we were running out to the fishing grounds, John said, "It's a miracle to have fishermen, trollers, seiners, and gillnetters, who catch the fish; tendermen who transport the fish on the collector boats; and shore workers who process the fish—with disparate and sometimes warring interests—in one union. That gives them the strength to shut down more than eighty percent of the salmon production on this coast. What keeps them together is the knowledge that without that combined bargaining power we'd all be right back where we were before there was a union: getting paid five cents a pound for salmon, and sometimes twenty-five cents or less for a whole fish. I have shares in the Prince Rupert Cooperative, and fished for them for many years, but I had a roaring fight with them during a strike several years ago because they continued to fish, and I will probably never fish for them again. I started fishing in 1934, and we used to get six cents a pound for coho, and up to twelve cents a pound for mild-cure, cleaned. That was for the Berlin and New York markets, where they brought the highest price. Twelve cents a pound for us, but smoked salmon would cost about a dollar-eighty or two dollars a pound in the market. Before the mild-cure process of preserving them in soft brine was used, the large springs used to go out in great big salt barrels, and the importer would smoke and finish them himself. We couldn't even sell white springs, so we often gave them away, or got a cent a pound, sometimes two cents, for use as halibut bait. You didn't know whether you had red or white springs until you cleaned them. Seafood Products pays the same for both, but most places still pay less for white spring, because the public has this attitude that salmon has to be red." He put his hand to his throat. "I'm getting hoarse. I should stop talking," he said. He opened a bottle of beer, took a big swig.

We fished for several days between Watch Rock and the Pearl Rocks. Each night at 8 p.m. John turned on the radio to the fisherman's band and listened anxiously to union representatives reporting on negotiations. Our life had an undercurrent of uneasiness; we wondered what would happen next.

One night, there was a full moon, followed by a spell of rainy, chilling cold weather and westerly winds, so John was careful to anchor in a protected harbor under the lee of the land. Early the next morning, the weather was so stormy that we stayed in the harbor, and, after the news, John turned the dial of the Wesco to the international radio band. An American fishing boat running to Alaska was asking permission to go into Winter Harbour, a settlement with a government dock, a general store, and a post office, on the west coast of Vancouver Island. John always listened attentively to any news about Winter Harbour, because he had fished off there for so many years and owned a lot on the waterfront, where he occasionally suggested we should move.

Canada's Federal Fisheries patrol boat, the *Tanu*, answered the call and gave the boat permission to go into the much closer but uninhabited anchorage at Cox Island off Cape Scott, north of Winter Harbour. When the American boat's skipper said he didn't know where Cox Island was and that his splashboards had all been washed loose, the *Tanu*, with obvious reluctance, permitted the vessel to go to Winter Harbour. "You can come in for repairs if it is absolutely necessary, but you can only stay long enough to make them and must leave again immediately," the voice on the *Tanu* said sternly. "You know the regulations: keep your poles up and your gurdies off and in your fishhold, and be out again by twelve noon tomorrow, or get further permission."

"All those American guys want to do is get in and sleep, but if an issue is made of it the *Tanu* has to have a proper reason for letting them—maybe a broken shaft or mast," John said. "If you want to do something that's not exactly right, you should know the other guy's law and give a reason that will cover it. Your American fish-patrol boats do the same—especially in Alaska, with its hundred-mile winds—and when some of our fishermen have asked, your Coast Guard has said no, if there was any American boat still to come in. In a real blow, coming in can take nine, ten, or eleven hours, making maybe two miles an hour for forty-five miles."

Shortly after, another American boat called in and the *Tanu* read them the regulations again, giving them permission as well, with the same strict limitations. John switched back to the Mickey Mouse in time to hear someone describe a birthday cake he was making for a party he and a friend were planning in the harbor. Later, he went out and waved at another boat, a red troller, as it came past to anchor. "That's Walt Nygren and his two sons," John said. "He's only got one arm, and he's one of our best fishermen."

That day, since it was too rough to fish, I cleaned out the cupboard—John called it the locker—under the sink, a half shelf at a time, setting the bottles and bags, pots, and jars that contained staples like brown Demerara sugar, flour, coffee, tea, oil, vinegar, and numerous condiments over on the bunk while I wiped off the shelves. Especially in rough weather, the locker's contents shifted, no matter how tightly I had packed them against one another. After a week or so, the shelves would be in a terrible mess and need tidying again.

During the cleanup, the floor had to be kept clear as a passage for John. Each container had to be laid carefully on the bunk canvas so that it wouldn't roll when the boat heaved, which was all the time in the groundswell. I enjoyed fitting things back into the cupboard again, closely packed to brace one another in niches that were now as familiar to me as they were to John: the coffee can, far left front, bracing the Scotch whiskey bottle directly where the doors slid open, kept in place by the wide jar that held pieces of Britl-Tak, both supported by the battered big rectangular blue-and-white painted tin decorated with pretty pink flowers which held the pancake mix, next to the round can containing sugar, next to another of flour, and so on, right along the two shelves, upper and lower, into the far corners where we kept seldom-used items, like chopped nuts for baking and unopened jars of preserves and pickles. It was important not to change the location of any item; everything on the boat had its own special place. "It could be a matter of life and death to find something in a hurry by just reaching in for it," John warned, one day when he found I had used a screwdriver and thoughtlessly dropped it back on the tool shelf instead of in its proper slot in the wrench box. "Besides, looking for things is a huge waste of time."

The storm abated, the weather turned sunny, and we were fishing again. I could no longer bear to stay inside at my typewriter, and to justify abandoning it I washed the *MoreKelp*'s windows. Mary Gunderson's daughter, Vera, had told me in Port Hardy about washing windows with newspaper as the scrubber, and warm vinegar water in place of soap, so I put warm water and vinegar in the laundry pail, and shoved it and a newspaper through the wheelhouse window. Then I went out and crawled along the narrow deck on the starboard side to the bow, holding fast to the roof, and washed the front windows. (I later discovered that the vinegar-and-newspaper process was as common as using commercial Windex, but it was all new to me.) I was delighted with the results, which made me daring enough to wash the side windows, too, although the boat was moving, slowly. I crouched on the narrow planking, my rear half over the gunwale, feeling the wind in the empty space between me and the sea, seeing the foamy water rush by out of the corner of my eye, while I gripped the wall with one hand and washed each window with the other.

It was such a lovely, clear day that after lunch, and an extended nap that began during the Music Hour when he pulled up the gear and flopped down outside in the sun, John decided to take a bath on deck. He filled the kettle and the bucket with water at the sink, put them both on the stove, turned it up, and, while he was waiting for the water to warm, stripped all his clothes off. He leaned out of the pilothouse door and looked around. "All clear," he announced. "We've got the whole ocean to ourselves. I am going to celebrate this day by showing you how to have a *real* bath at sea."

He soaped himself from the top of his head to the soles of his feet, grabbed the bucket, and stalked out on deck, where he dumped the whole bucket of water enthusiastically over his head. I hastily took up a position on the far corner of the hatch lid to escape being splashed. Standing on the deck naked, with water dripping from his head and beard, he had a kind of walrus look, and I began to laugh. "I've never seen anybody look so *wet*!" I said. I was going to offer to get rinse water when he rushed back into the pilothouse, poured the contents of the kettle into the pail, hurriedly pumped water in from the sink to fill it up, ran back out on deck, and dumped the contents over his head again with the same abandon.

We had had such a pleasant day and John had such a long nap in the afternoon that I announced a cocktail hour when we anchored, and put on the one long dress I had brought with me in the bottom of my pack. Before I came, I had the idea that from time to time we would go into charming little seaports like the ones I had seen at Cape Cod in Massachusetts, or around Chesapeake Bay in Maryland, and step out for dinner to the small, chic kind of restaurant one finds in summer along the Eastern seaboard of the United States. It was already clear to me that, living with John, I would be saying goodbye to all restaurants except those that were practical, cheap, and preferably Chinese, and that there was nothing I needed less than a long, low-necked dinner gown. The dress was bright-red cotton—made specially for me from cloth sent by friends working for the Canadian government in Lesotho, Africa—and I was determined to wear it just once. To my surprise, John was delighted with the idea, put on a clean blue shirt, and opened a split of champagne that he had hidden under my bunk. I was curled up on the bunk, laughing, my bare feet tucked beneath me, when John suddenly swooped from his perch on the steering seat, seized the square mirror from its slot behind the sink, and held it up to my face. "Look at yourself!" he exclaimed. "Look at yourself in the mirror and tell me whether you think this fishing life with me agrees with you or not."

I looked from his beaming face into the mirror and saw, as if for the first time, the sunburned face, brown eyes, white hair, and freckles of a woman I was familiar with, all right, but the glowing look was new. I turned back with astonishment to John, who was still holding the mirror up to my face. "Is that really me?" I said, and covered my face with my hands. He took my hands away and looked at me.

"I guess I've come to stay," I said.

XIII

The next morning, we were up at six and fishing back and forth between Pearl and Watch Rocks in wild surf, with many gulls following our boat. The fishing was wonderful; John caught four huge springs while I sat on the deck beside the fish box and watched. We were coming once again south-southwest, around behind the breakers at Pearl Rocks. The sun was high in the sky, and its rays shone straight down into the water.

Something felt different. The *MoreKelp* was surging forward, and the rumbling of the Gardner engine took on a heavier note. We're going too fast, I thought—much too fast for trolling. I leaned over the side to see why. Jagged peaks in the water below us were so clear that I gasped. Just then, the Gardner seemed to die: my heart stopped. Terrified by the stillness, I looked up just in time to see all the fishing lines drop straight down into the tide boiling over the rock piles into which we were heading. What was wrong with the Gardner? Why had it stopped in this awful place? I stared at John. His face had that intensity that it got whenever he was fishing dangerously. He had one hand on the wheel while he looked over the side at his lines. The amazing thing, I thought, was: we are going to pile up on the rocks and John doesn't seem to care. He straightened up again, looking—it was hard to believe—blissfully happy.

The quiet—so very quiet—lasted an eternity. The boat swayed, pushed by the swirling currents, and I thought: Here we go!

Was that sound the Gardner speeding up? Yes! I could hear the engine's rumble over the roar of the surf. Thank God! I was standing up now, clasping the chain rail, staring at the breaker, expecting to crash. I looked around at John. He was grinning as he made a slow turn to starboard, into quieter water. I looked up at the port pole tip; its spring was moving violently. We must have caught a fish. He shouted, "Did that scare you? Hold tight. We're going to do it again!" And he did; and once again after that. Each time, the sound of the Gardner was reduced to a whisper. Each time, I found it hard to breathe until I heard it start to grumble again.

"That's it. The tide has changed again," he said finally. We continued north along the west coast of Calvert Island while for the second time that morning he was washing gorgeous huge springs, all mild-cure, that he had caught. I sat down on the edge of the fish box and said, "Please. Please explain to me what you were doing, so I don't die of fright the next time the engine nearly stops in the middle of fishing so close to rocks. I thought we were in real trouble."

"I can only make this maneuver if the tide is right, on the flood for Pearl Rocks, when the sun is shining right down into the water and I can see those horrible bloody peaks clearly," he said, and laughed. "It's hair-raising all right, but I catch my biggest springs when the tide eddies behind the rocks like that. What I did: I turned and headed for two landmark peaks with the sounder showing twelve fathoms, and I kept opening up and opening up the poor old Gardner until we were going six knots, putting on that speed to keep the gear up," he explained. "Then I suddenly throttled the engine down to dead slow so the gear could go straight to the bottom and hang behind the rock, where the fish were lying in the eddy. From their point of vision, the lines are surging forward like living food, and the springs start up and follow the lures and strike at them. It's a way of getting lines down into holes and crevasses, and a wonderful trick, as long as the motor is dependable. Otherwise, you're on the rocks, and the gear is tangled and full of junk." He leaned over and patted my hand, which was gripping the guy wires. "You have to know what you're doing," he said. "So far, the Gardner has been good to me."

John had been predicting a southeasterly gale ever since I washed the windows, and now the sky was overcast. "Fishermen are like farmers—always looking at the sky in relation to the land to see the speed and direction the clouds are moving from," he said. "We watch them all the time for any change. You can see the top clouds going like hell, and that means it's blowing upstairs and eventually going to come down. If you see a black streak coming along fast—sometimes as fast as thirty-five miles an hour—go find someplace to hide. I watch the weather *all* the time."

He threw out the gear again at a place called Dublin Point with the remark "We sometimes get a spring here." He caught another twelve salmon—a mixture of smaller springs and coho. The sky was filled with threatening black clouds, so we ran into a good harbor.

Early the next morning, we heard on the radio that the fishing strike had started. For once, we actually sat down to a leisurely breakfast of scrambled eggs, bacon, toast, and coffee, before we went through Hakai Passage to Namu. I was washing dishes when an enormous groundswell sent the frying pan on the counter crashing to the floor. Meanwhile, John was heating water in the pail on the stove to wash down the hatch after he delivered his fish. "I don't like B.C. Packers, but I certainly am grateful for Namu, because it has helped so much to keep the fleet fishing," he said. "If you have to have something fixed, like a radio or a winch, they'll have you in and out in a day there, rather than several days to a week someplace else. It's a far better and cheaper spot to be broken down, with a machine shop at the float. Namu has many quiet floats and a few horrible lights, while Hardy is now a *blaze* of mercury gelo lights. I sure do like Port Hardy best for selling my fish. Both the Hardy and Namu plants are union, but Seafood Products at Port Hardy is a small company and a loverly atmosphere, plus a good employee situation. On the other hand, there's no logging at Namu, no road, and no winter population; also no appalling marine traffic jams. It's easy and convenient to work out of Namu, and social life is a blank compared to Port Hardy, which means that for me it is one hundred percent less exhausting because there's *room*. There's no choice to shop, a lousy company store, and no liquor store, but there's *room*!"

We arrived at Namu at 10:30 a.m. to sell the fish we had caught before the strike. We were the second-to-last boat in. The whole fleet must be here, I thought; I had never seen so many fishing boats tied up in one place, five and six deep. The tops of the trollers' upright paired poles made a fine design of thin vertical lines above float after float; the air was thick with them. In among the trollers, which I would naturally notice first, I slowly began to pick out massive drums with nets wrapped around them and high curved decks of towering seiners. Even though I disliked seiners because they swept up all the fish in whatever area they set their nets, in my pre-*MoreKelp* days they were my idea of what fishing vessels should look like. I loved the majesty of their lines and size. They were so big that they dwarfed neighboring gillnetters, whose smaller drums were like toys beside them.

John got his southeasterly gale as we were turning to go to the cannery. The clouds opened, and rain streamed down on us as we tied up at the unloading dock—a wall of concrete from where I stood. A wire ladder was swinging from the top; it could not have threatened me more if it had been a grizzly bear. A head appeared above it over the side of the dock, and a young Chinese shore worker called down to John, "Do you want me to put your name down on the board for the lineup? You'll be number 11."

John shouted back, "Yes, thanks." As he pulled off his wet Stanfield's and hung it over the stove, he said to me, "We wouldn't be able to fish anyway in this weather."

We sat down in the pilothouse to await our turn to unload, and a large black bird arrived outside on the gunwale and stared at us. John took a wooden whistle from the shelf over the sink, leaned out the window, and blew it, producing a cawing noise—"Gaak gaak." The bird looked around, and listened intently as John continued to make crow calls, watching him until it must have decided he was some kind of a crow, too, responded with a firm "Gaak" of its own, and flew off. John put the crow call away and said, "I have to stay here to move the boat, in case our turn comes. Will you run the mail up to the main office while we're waiting here? One of the pilots from the small airline that services Namu will be coming in on his regular schedule and can take it with him."

"How do I get up there?" I asked nervously. But I knew the answer. I had already counted the rungs on that ghastly ladder between me and the top of the dock. There were eight.

When I lived with John, the first attempt at anything unfamiliar was like stepping off a high diving board. I usually improved with repetition—except with swinging ladders. My trusty new ladder had three steps and lay firmly against any place to which it could be attached. The long ladders at the fish-plant docks at low tide were dangling shivery structures; as I looked up at them from below, they disappeared into the sky. Fish docks on concrete pediments or creosoted pilings seemed to be especially high above the water and were weighted down with ponderous derricks, winches, and heavy ice-making machines. All this weight probably required the reinforcement of the heavy beam edgings that stuck out. When I climbed up or down their ladders, I had to swing out in space around these beams at the very top and scramble over them without being able to look at my feet. All I could do was pray that they would land on the next rung.

I can still recite the list of B.C. dock ladders that scared me out of my mind: at the government floats in Port Hardy, on Texada Island, and at Namu. The worst one, which plagued me all our life together, was below the ice chute at the Campbell Avenue fish dock in Vancouver. In the spring and fall, when John fished near home and made a six-to-ten-hour trip (depending on the weather) in the boat to Vancouver to sell his fish, we would go to Campbell Avenue dock, unload our fish below the headquarters of John's favorite fish buyer, Norman Johncox, of Billingsgate & Co., and then move under the ice chute to be first boat to take on ice when it opened at five the next morning. We would set off for home as the sun rose, gliding in the quiet of the early morning under the graceful span of the Lions Gate suspension bridge, which connects the municipalities of West Vancouver and North Vancouver to the business center of the city of Vancouver, and on into the open water of Georgia Strait, crossing Howe Sound, with its rim of snowcapped mountains off the starboard bow.

The first time we went to the Campbell Avenue dock together, after we had finished selling our fish, John wanted to take me out for dinner to his favorite restaurant, the On-On, in Vancouver's Chinatown. I stood on the deck of the *MoreKelp* facing the black oily pilings, looked up the thirty or forty feet—it might as well have been a hundred—of ladder swinging from the overhanging ledge of the dock above, looked down at the garbage floating in the water below, and said to John, "No. You go out to dinner and tell me about it. Or bring me some."

"Nonsense!" John said, disappearing into the pilothouse and coming back with a coil of heavy rope in his hand. He tied one end around me under my arms. Then he ran up the ladder, paying out rope along the way. At the top, he stood on the pier, holding his end, and instructed me to start climbing. "I'm right here holding on to you," he called down. "Nothing can happen to you, with me at the other end." I felt like a dog on a leash, but up I went. When we returned home that night, he tied my leash around me again, stood at the top, helped me get on the ladder over the terrible edge, and slowly payed out the rope as I descended, so that if I fell, as I told him later, we would both go into the water. If anyone else was awake when we returned to our boat late at night after a social evening in Vancouver, we must have made quite an impression: me descending on my leash, John standing at the top, paying out rope and encouragement.

At the Namu dock, John ran me up the ladder on my leash and I delivered our mail, wrapped in the usual plastic vegetable bag, to B.C. Packer's front office. I stopped on my way back at the company store to buy some groceries: a weary head of lettuce, two bananas, eggs, and milk—as little as possible when I saw the food prices.

As I came out on the dock, a handsome young woman with red hair in braids wrapped around her head ran up the ladder—how I envied her!—and stepped out on the top of the dock not far from me. I had noticed her earlier on the deck of the troller called *The Venture*, just ahead of us in the lineup, which was now unloading its fish. Her husband was standing below beside their open hatch, whose huge cover was laid back on the deck, guiding the pails that swung down on the winch into his hold, where a young boy was loading them up with fish. She ran over to the sorting table as the fish were brought up and dumped there from the pails, and then followed them to the weighing platform, making notations on a small pad she was carrying. I was impressed. I thought: If I am going to be of any use to John, I had better do this, too. John, in his oilskin rain gear—waterproof overalls, coat, and hip boots—was busy down in the hold throwing out the fish blankets on the deck. "If she can climb up to this dock, I can go down from it," I muttered. With my heart beating right up in my throat, I moved slowly down the ladder, without my leash, my grocery bag swinging from my arm. Hand over hand, gripping the wire sides of the ladder so tightly that my knuckles were white, I descended, feet groping from space to space, until they set down, as if they were independent of my head, on the solid planking of the float beside our boats. I climbed into our boat, dropped my groceries into the sink, and ran over to the open hatch. "I'm going up to watch over your fish," I yelled down to John. I climbed up the ladder and learned something about myself: If I absolutely had to go up or down a ladder without John and the rope, I could. Exception: Campbell Avenue.

When our turn came to unload, I looked over the side of the dock at John, who had stopped scrubbing pen boards to move the boat forward. The young boy came aboard and began loading the descending buckets with fish from our hold. When the first bucket was hoisted up, overflowing with our salmon, which hung over the edges haphazardly, a youth at my level above grabbed it, opened a latch at the bottom of the bucket, and slid the fish out onto a wooden-slatted platform, sorting the salmon into species and dumping them into square metal wagons big enough to hold seven hundred pounds.

The rain was coming down in such a torrent that I could barely see through it, dripping down my face and off the end of my nose. To escape the downpour, I walked into the shed behind the wagons holding our fish, and stayed while they were being graded. A young man with a red beard was standing against the wall, observing, so I went over beside him. He was a biologist from the University of British Columbia, working for the summer as a fish sampler in a hatchery recovery program. "As the salmon are brought up, I watch for salmon tags, either an internal nose or adipose-fin clip. If the salmon has that tag, we want that information for Federal Fisheries," he explained. "It's part of a ten-year study that takes in the area from Prince Rupert to California, to show the intertwining of American and Canadian fish: who catches the fish, and the distribution. We can tell the number of fish Canada is catching and vice versa. The tags give the brood year, whether it was a fall or spring run, where the fish was let go, how old it was, and its size. All five Pacific salmon species are a little different; they spawn in fresh water but live in salt later, and some spend a year in a stream, some in lakes, some in the ocean. You can even tell the rate of growth from the scale samples we take; they are like the rings on a tree."

When I went outside again, the rain had stopped. I looked over the edge of the dock. John was just pulling away to let in another troller. I saw the *MoreKelp* go out of the unloading area, and then I watched the distinctive diamond-shaped metal piece at the top of our port trolling pole travel around through the thicket of stationary poles until it stopped moving. I had come back too late to go down the ladder; thank God. I set out on land in the general direction the boat had taken, over to the float area, where all the fishing vessels were tied up, and walked up and down the maze of floats until I saw John. He was leaning over the chain rail of our boat, which was on the outside, talking to a shorter man whose vessel was between us and the float. The other man looked so much like John Chambers—except that he didn't wear glasses, he had less hair and what he had was whiter—that I wasn't surprised when John introduced me to Chambers's younger brother, Jimmy.

"We're just leaving for the union meeting. Come along," John said, so I followed them from the float along a path to a marine ways. Men and women of all ages, sober-faced and quiet, were sitting on top of and beside a gillnetter up on the ways for repairs, and on the rails used to pull up boats. The dampness of our surroundings, the grayness of the day where only a thread of yellow peeping out from the dark clouds across the sky hinted at the existence of a sun, accentuated the gloom of the occasion; all present would lose a sizable part of their year's income in a strike of any length at the height of the fishing season. I sat in a kind of daze at this twist in our fishing life—unexpected by me but apparently a relatively common occurrence. A steering committee for Namu was elected, and contract terms were discussed. "No one-year contracts; let's make it good enough to last two years," someone said, and there was a buzz of approval. I was barely listening, recalling something I had heard on the float, on my way to find John. I had passed two young fishermen, and one was saying, "If the strike goes on, I don't know what I'll do about the payments I owe on my boat. I can't get any unemployment insurance, I have a wife and kids, and there won't be any jobs. God, I stopped drinking, and now this!"

At the end of the meeting, on the way back to our boats, I listened to John and Jimmy talking about exhaust pipes. I was surprised again and again at the intricacies of fishing revealed in ordinary conversation; and exhaust pipes, which I had never thought about at all, were no exception.

"My exhaust pipe used to rattle, so that I could never catch any fish," Jimmy said.

John replied, "I've always had mine hanging from the inside with big springs."

"Well, that's what I'm doing now, and I'm fishing real well," Jimmy said.

When we arrived back at Jimmy's boat, to which we were tied, John said, "We call these floats the Toonerville Docks. Probably the name came from a newspaper comic strip in which everyone sat around and did a lot of talking." Jimmy's troller, the *Kitty D.*, had a small wheelhouse abovedeck—big enough for one person to sit down in and from which Jimmy steered. The wheelhouse contained the VHF and Mickey Mouse radios and a sounder; switches for all the wiring behind the seat, including a light for the engine room below; and an instrument panel that showed all the engine gauges. His living quarters were in a trunk cabin underneath. "All the fish boats used to have sleeping and eating quarters below like this," John explained. "The *MoreKelp* was the first fishing boat in our harbor with a big pilothouse containing all that you needed abovedeck, and everyone came around to see it. I find it much better when I'm fishing alone—especially when I have to get up suddenly to move the boat, or in any kind of bad weather. When I slept below, I could never get proper rest. I was always worrying about what I couldn't see that might be happening outside."

"I've never known anything else but this, so I'm quite comfortable with it," Jimmy said as he went ahead of me down three small carpeted steps into his cabin. It had simulated-brick linoleum on the floor, a little wooden table that folded flat against the wall, and, behind it, eating utensils, glasses, jelly, spices, and other staples. The bench I sat down on had storage lockers underneath, and I faced a small stove whose oven had an indicator that I envied, since I had to guess at the heat of ours. Jimmy's stove was under the wheelhouse close enough to where he steered for him to reach down to it from his seat. There was a little sink with a tiny pump, and towels and dishrags hung from green-and-white painted racks attached to the ceiling. Jimmy's pocket watch was fastened around another rack on the wall.

At the bow were two bunks, one coming from either side; whoever slept in the lower one would have his feet right under the other bunk, with only inches to spare. I thought, I would stub a toe every time I kicked off the covers, and appreciated my own bunk as I never had before. I heard music and discovered a radio on the far side of the stove, toward the stern, along with batteries, fuel tanks, and the engine, which was below the wheelhouse. I knew there was a small closet with a toilet, or head, tucked away somewhere, although I didn't see it. Portholes along the sides, a skylight in addition to a ceiling light, a tiny electric fan, and the curved shape of the cabin, which occupied the whole forward area below, somehow blended with the other furnishings to make a snug atmosphere. Our pilothouse seemed to lack that homeyness, perhaps due to its long, skinny shape—more probably because there was no truly comfortable place to sit down.

Fishermen in Namu could have their meals at the cookhouse for a small fee during the short thrice-daily period when it was open. John said the meals were excellent, and it would be a change from cooking on the boat. We invited Jimmy to join us, and we all climbed over into the *MoreKelp* to have a drink outside while we waited for the supper hour. John unfolded the deck chair and sat down in it, Jimmy settled on the hatch cover, and I on my foam-rubber cushion to enjoy the late-afternoon sun, which had unexpectedly come out. The men talked about the strike, and I listened.

When we came back from dinner, John untied our lines from the *Kitty D.* and we left. "I can't stand the noise and all those other boats," he said. "We're going to Boom Bay."

It was like hopping in a car to go around the corner. John had barely started the motor and moved out into open water when we were going through a narrow opening between rocks and trees into a small inlet—Boom Bay.

While John tied up to a decaying half-sunken dock, I looked around. We were parked in a graveyard of old floats and fish-camp buildings that John said B.C. Packers was storing along the beach around the bay. The background, trees high enough to melt mistily into the lowering clouds, was a deep blue-green, and the water around us was very black. Fallen logs littered the shore and floated around in the murky water. The graying light vanished without giving us a glimpse of anything so cheerful as sun or even a few of its rays.

John came in, and I was about to tell him this was one of the most depressing places I had ever been, when he said, "Namu is not really a harbor. It's too open. The word Namu means 'whirlwind' in the language of the neighboring Bella Bella Indians. It blows so in winter in Namu and there is so much wave action that B.C. Packers moves almost all the floats from there to here and doesn't put them back where you saw them until spring. I've been coming to Boom Bay to anchor for years. Our own name for it was Bachelors' Bay because of the number of bachelors, before the days of unemployment insurance, who lived here in float houses—regular houses built on floats that can be towed from place to place. They would fish in summer, go to Vancouver in the fall, and around Christmastime they would come back and do their spring overhaul of their boats. In those days, it was a cheap place to live. There were always from two to five float houses here, and lots of crabs to eat."

John poured himself the customary shot of Scotch that marked the shift from his preoccupation with the day's work to his preparation for dinner and bedtime. "Oh, this place brings back memories," he said, sitting down on the steering seat. "Very few people know about Boom Bay now, but in the late thirties they were all here: Bob Merkle, Alec

Gow, Walter Wright. If you needed to do last-minute rigging up and work on your poles, they always had a planing bench; one had a drill press, another had a band saw, and among the three of them they had a wonderful collection of old pieces of pipe, hose, and generally useful junk that all fishermen need. As a youth, Bob Merkle had drifted down the Mississippi River with another kid, and they used to steal chickens that a black woman would cook for them. He used to say, 'Boy oh boy, *that* was chicken! I'm talking about *chicken*, not these goddam frozen seagullies we get now that are not fit to eat,' and he'd roll his eyes back so far the pupils would disappear. He also had trapped all through the Northwest Territories, and once he told me he had gotten a thousand dollars' worth of furs from an Indian on condition that he come back and marry his daughter." John pulled the stool down and stretched his legs across it. "Bob said, 'I never did get back, but I always wondered what she was like.' Then there was Walter Wright, who had a little gillnetter. That was probably not his real name, because he was French. I *loved* his French accent. He had huge eyebrows, always wore a woolly toque, smoked a huge meerschaum pipe, and was a wonderful cook. On harbor days, he would call to me and I would row over. There would be a marvelous smell of cooking, and he would be drinking gin and making pies. And Alec Gow. He was an extremely kind and cheerful man. Two old people, Joe and Agnes, lived in a float house near here; they were a real love story. Joe was Irish and Agnes was a Rivers Inlet Indian. They fished, and she taught Joe to trap. Alec was always taking their mail to them, and their old-age pension checks. When Agnes became very ill, and Joe was exhausted from taking care of her, Alec took them to hospital. He brought them home again and they toughed it out on the float another four months, until Agnes died. Then someone towed Joe and his float house down here. His son flew up and tried to get Joe to leave with him, but he wouldn't. The manager of the cannery used to pop in, but he was busy, and Alec looked in on him, but he'd have to go again. Joe lasted only two weeks after Agnes; he died of grief and exhaustion."

"And Alec?" I asked.

"All, all gone," John said. "Alec was fishing and lay over at Goose Island for a week in a sou'easter one August, and had a very bad stroke. He had a friend with him who got on his phone to the other boats at Spider Island camp nearby, and they sent an airplane for him, but it was very rough and Alec was so heavy they couldn't get him into the plane or even on the pontoon. They put him in his bunk on his boat and towed him back to Spider, where they did manage to get him on a plane. He lasted only three days, fortunately. Lack of exercise on the boat, I think, did it."

XIV

In the morning, we had breakfast outside, John in his deck chair, with his plate on his lap, and I on top of the hatch, my back resting against the mast. We still were tied up at the ruined dock, surrounded by old floats, battered oil tanks, and boats left on the beach to disintegrate—all bathed in a kindly mist. The enclosing shores screened us from the world outside. The stillness was broken only by an occasional crow cawing overhead or the haunting cry of a loon. Like John, I had fallen under the spell of Boom Bay. I looked at his gray head, bent over the task of spooning a soft-boiled egg from its shell in a glass egg cup. In the shadows of the black water, a dark pool edged with tall trees down to the shore that shut out the morning light, I tried to visualize the lanky, dark-haired, clean-shaven young man I would never know except from photographs. "Hello, stranger," I said to him. "I am imagining you thirty years ago, rowing over to eat pie with the Frenchman. Tell me some more."

"What would you like to know?" John asked.

"I guess I'd like to start at the beginning," I said. "With your parents and where you were born, and so on."

"My dad's full name was Joseph Heywood Daly, and there were MacGillycuddy cousins who had big estates in Ireland," John said. "There were also some Quayle relatives in Ireland. I'm English on the Broadbent side, my mother's family, but my political sympathies are *all* with the Irish. God, those cold, wet beds in English houses! They were almost the death of me. It's the one thing I can never forgive my father for: letting my mother send me to those terrible English schools."

He threw his eggshell over the side, got up, and went into the pilothouse, returning with two pieces of toast, the jar of marmalade, and more coffee. He sat down and continued, "Horse racing was a popular sport on the Isle of Man, and as a young man my father won many races. After his own father died, my dad's mother married a German and took Heywood with her to Germany, which he hated. He was sent to a Prussian military school, and he ran away. His mother then let him emigrate to the United States, to an uncle who was a river pilot in Oregon between Portland and Astoria. At seventeen, my dad contracted tuberculosis, and someone told him that the air in Montana might save his life—so in 1896 he packed up and went there. He became an American citizen, always wore a Stetson hat, and spent the rest of his life in Montana, except for two years tending cattle in Argentina. When he got T.B., he wasn't given long to live, and he died in Billings, Montana, at the age of ninety-seven. If you picked a map with no town, no nothing, my dad would be in the middle of that. He was a real roarer against big cities all his life, believing, and rightly, that they breed misery and *do not produce food*, so I come by the feeling honestly that big cities and airports are mankind's cardinal sins committed against those glorious gale-swept mornings and sunsets you've experienced with me already. Honestly, dear, as I get older, just the thought of cities makes me shiver and shake. My dad knew. 'Cities will be too much for you,' he said. 'We are outdoor-equipped—for outdoor lives, not cities.' "

John was silent for a moment. Then he said, "I remember my dad's old Model T Ford. If it didn't have enough gas to start, he left it backed up a hill so that it would be facing downhill to start running, and the gas tank would be higher than the carburetor and feed down to it. There was no speed limit in Montana, and he was eighty-nine the last drive I ever had with him, and it was sure time he stopped. He always had ranches and cattle, and he thought it was bad luck to be anyplace but in Montana. My parents' marriage was a great mismatch. I grew up on Vancouver Island, but I used to go and visit my dad and he would come to see us in the winter at Cowichan Bay. Not for long, because he didn't like to leave his cattle. One of my earliest memories is of throwing a new ax down a well on the ranch when I must have been about two and getting walloped for it. I don't know when my mother and I moved to Vancouver Island, or how many summers I spent with my father on his ranch in Montana. After my mother's death, I spent a whole winter with my dad on his Eight Point

Ranch, in below-zero temperatures, crossing on ice over rivers in his car, and 'punching' cows the modern way—driving across fields right up to the steer or cow and inspecting it without dismounting from the limousine. My first chore every morning was to go to the river and break ice for the cattle to drink, when there wouldn't be a breath of wind, the trees were all glistening with frost, and I could hear calves being weaned bawling above, on high ground. One of the things I learned to do, being with my dad on the ranch, was to cook. He used to say to me, 'You'll always be welcome anywhere if you can cook,' and I have found that to be true."

"You are a very good cook," I said.

He nodded. "I know," he said. "I'm not much on cakes, though. I used to make something called the One-Two-Three-Four Cake for my kids. Can't compare to that lemon cake of yours."

He went inside and came out again holding the open pot in which he made coffee, and poured us each a cup. "I'll never forget the first time I really rode a horse. My dad had put me up on one two or three times for an hour or two, and then when I was twelve, he said, 'I haven't had time to teach you to ride, but if you want to come out and round up cattle with me we'll leave tomorrow. *I did not know* what that would be like, which was *lucky*! We got up at three-thirty in the morning, had coffee and hotcakes, and were in the saddle by five. We rode until 11 a.m. without stopping, except to get water for us and our horses. We stopped to eat a sandwich and rode off again, with a few breaks to pick up cows, until it was dusk. Around 8 p.m., we staggered onto somebody else's ranch and were given beans and pork and turnip. We slept together in one bed, and rats and mice flew around all night, not to mention a few bedbugs. We were up at four, had a breakfast of salt pork about two inches square, corn grits, and coffee; were in the saddle by five-thirty; and rode again until dark, with two afternoon breaks of fifteen minutes each. The third night, we arrived back at our ranch about 5 p.m. with twelve head of my dad's stray cattle. All the skin was off the insides of my legs. My legs felt like fire and full of cramps all night, but my dad gave me *no* sympathy. He said, 'You did O.K.,' and kicked me out of bed to get the fire going for breakfast at five-thirty the next morning. He said, 'You can get as drunk as you like and go to bed when you like, but *never* be late for work, regardless of your condition.' He told me that in those old cow outfits he used to work for that was the unwritten law—no sympathy. You could play poker all night, but you were in the saddle at five or five-thirty the next morning. Because of his T.B. background, he rarely played poker. He went to bed, instead. In height, he was barely six feet, and he always said to me, 'You are tall, so be careful, because at the cow outfits the tall men wear out much faster when sleep is short than the short ones do.' I found that to be a hundred percent correct, in my case. The 4-Z ranch in the Crazy Mountains near Melville, Montana, was my dad's last ranch, and he had that until he was well into his eighties. I was very fond of his second wife, Belle, who came from Victoria. She died before he did. In his nineties, he still enjoyed stepping out with the ladies and liked to dance. He went to Hawaii with an old pal who was the proprietor of a hotel in Juneau, Alaska—widow chasing, they said—but after two days in Hawaii he wired us that he was going back to Montana immediately, and he never left it again."

"What was your mother like?"

John was spreading marmalade on the toast, and he stopped, knife in midair. "I have an oil portrait of her at home as a young girl," he said. "She was lovely-looking, with dark eyes and long, brown hair that turned white very early, like yours. There's a family story that she rolled her eyes at an auction and a big brass bed was knocked down to her. She was very spirited. Everything about nature thrilled her. We would be driving down the road at Cowichan Bay in our Model T and she would zigzag across it without a thought about oncoming traffic and say, 'Look at that maple tree, John. Isn't it beautiful?' My mother loved the outdoors and was a fine ice skater. She was highly critical of the way I skated. I enjoy skating, but I find skiing far more satisfying."

He finished putting marmalade on the toast, and handed a piece to me. He took a bite of his, crossed his legs, and continued, "We had wonderful times together. I remember long hikes, and fishing in my little boat in Cowichan Bay. My parents were married twelve years before I was born, and I was a shy only child, brought up in lonely areas of Montana and Vancouver Island. I had very few playmates—really only one boy my age, the son of a family friend. I try to be easy in a group or a party gathering but am usually most nervous underneath and talk far too much, to cover it up. I was the first grandchild, thoroughly spoiled by my doting grandmother, who lived with my mother and me at the house at Cherry Point. Ruth Chambers, who is my oldest friend and has a farm inside the city limits of Victoria, where she and her husband, Lawrence, raise Brussels sprouts and potatoes, remembers coming with her mother to visit at Cherry Point when we were very small, and we were told to go down and play on the beach. Every time she sat on a log, she tells me, I would say, 'Don't sit down on that log; it's *my* log and this is *my* beach.' "

He got up and went in the pilothouse, returning with the mirror and scissors, and continued to talk while he trimmed his beard and mustache. "My father became very American—even talked like one, despite being a Manx man, born and bred. He went back to the Isle of Man in 1903, when he was twenty-six, and sold the Heywood family seat, Bemahague House, to the Manx government. It is now the Isle of Man's Government House. He was there when the Hunt Club Races at Peel were on, that July. He saved the Manx newspaper accounts of his trip in a ledger that I inherited, and they reported that in the Members Race 'Daly fell just at the start but was not hurt and won easily,' coming in third. Later that summer, at an Agricultural Society horse show, he was judged best rider in the field. I also have an ancient parchment scroll at home that traces the Heywood family back to the twelfth century." John laid down the scissors and peered at himself in the mirror, turning his chin from side to side, and smoothed down his mustache. "Do you still want to hear more?"

I nodded, and he continued, "My mother's family lived in Cheshire, England, near the city of Chester. Her father, Alfred Broadbent, worked for the railways, and died in his early thirties of septicemia from a wound, leaving my grandmother with very little money and three small daughters, one of whom died in childhood from typhoid. My mother, Muriel, was the eldest, and my second boat, the *Muriel D.*, was named after her. A chemist in the Broadbent family invented Hudson's Soap and made a fortune, so when my Grandfather Broadbent died, the family gave his widow an income of five hundred pounds a year and a house to live in rent-free, called Bache Cottage, on the estate, where my mother and her sister, Dot, grew up. My Aunt Dot was highly educated; she graduated from Cambridge University, but neither Oxford nor Cambridge gave degrees to women in 1900, so Trinity College in Dublin offered all female graduates from either university the Trinity master's degree, for a hefty fee; I think it was twenty guineas. That was a *lot* of money then, but that's how Dot had her master's degree from Trinity College in Dublin instead of Cambridge."

John pulled off his Stanfield's as a thin ray of sun leaked through the heavy tree growth. "When my mother became pregnant, my grandmother wanted to make bloody sure that her first grandchild would be a British subject and be born in Canada, not Montana. She left Bache Cottage and moved to Canada lock, stock, and barrel, bringing her younger daughter, Dot, with her, but I don't think she ever really made her peace with Canada. She looked down on Canada, compared to England; so did Dot, and probably my mother did, too. My aunt quickly found a job teaching in a private girls' school in Vancouver. My mother came on from Montana and I was born at the Vancouver General Hospital in 1912, on Halloween—I always celebrate my birthday with a tee-rific bonfire. You'll love it!" He looked at his watch. "We should be going soon," he said. "I ought to find out what's happening about the strike, and I want Fred Welland, the electronics man, to take a look at my recorder-sounder. He's one of the best reasons for coming to Namu, but he's so busy you have to make an appointment."

I was puzzled. "Where does living at Cowichan Bay fit in?" I asked.

"Dot married Tom Stanier, who was the first physician to practice radiology in Victoria. He built a country house by the sea thirty miles away at Cowichan Bay, and in 1921 he gave up his Victoria practice and moved his family to the place at Cherry Point, where he and Dot each had their own dog, baked their own bread, and conversed with each other from separate bedrooms by yelling across the kitchen. They had two children—my cousins Roger and Diana. Roger, the older, is four years younger than I am, and one of the world's foremost microbiologists; his wife, Germaine, is French and a microbiologist as well, and they both work at the Pasteur Institute in Paris. Diana is married to Ted Knowles, and the Stanier house at Cowichan Bay is now theirs. My grandmother purchased a place two doors away from Tom and Dot, and I came with my mother to live with her there at Cowichan Bay."

"What was the house like that you grew up in?" I asked.

John had been walking around in his heavy socks, and he fetched his shoes and sat down again to put them on. "Oh, it was full of periodicals and newspapers from 'home'; that is, from England," he replied. "It was typically English, with a veranda, silver picture frames, and cretonne curtains and couches, and the MacGillycuddy Sword—it's in a trunk in my basement now—hung in a scabbard on the wall in the living room. My father sold that house right after my mother's death and auctioned the contents, including Granny Broadbent's things, without telling the family, which didn't go down well. He wanted to get back to Montana, I guess. My grandmother was a typical Victorian bluestocking matriarch, used to having her way, and she and my father never got along, even on his short trips to see us in the winter. Roger and Diana were younger than me by several years, so we didn't have a great deal to do with one another when we were growing up, but we are close now, and I like them very much. We'll make my annual trip to Vancouver Island in the fall to visit Diana and Ruth Chambers. Roger comes to Canada every summer—to a cottage he built on one of the Gulf Islands—and I usually try to go by on the boat and see him then."

John suddenly bellowed, "God, how I *hated* those bloody English schools I was sent to!" I jumped at the violence of his anger. He lowered his voice, but not much. "It was apparently unimaginable to send me to the one-room elementary Bench School three miles away on Cowichan Bay Road—we weren't allowed to mix with the so-called common kids—and Ruth Chambers once told me that her mother didn't even want her to go to the public library because the books might be dirty and full of germs. At the tender age of nine, I was sent to board at St. Michael's Private School for Boys, in Victoria, and, as if that weren't bad enough, at fourteen I was packed off to England, to an English public school called Sedbergh, a private school in the north of England that my family really couldn't afford, where the sun never shone and I spent four horrible homesick years. You've never seen anything as bleak as those Yorkshire moors, and it rained all the time. During vacations, I used to visit three spinster Broadbent cousins: Bay, who was a justice of the peace; Lucy, who wrote a travel book called *Under the Italian Alps*; and Sylvia, who had very large feet and a mustache. I used to take long walks with her on the moors. They lived in a big house called The Hollies, which had been a rural estate but was surrounded by then by industrial Manchester. They were poor as churchmice, but they insisted on continuing to live in some kind of ghastly grand style, which required them to dress for dinner every night in appalling evening dresses that exposed their skinny bare shoulders, in a house that was absolutely freezing, served by the garden boy in a white jacket with brass buttons with the Broadbent crest, which was a flaming torch inside the V of a broad arrow. The bloody English gentry! I loathe everything they stood for! All that b.s. about social inferiors and their own superiority; all that emphasis on appearance and behavior, on military careers, and the fake religiosity! Years later, my Aunt Dot was talking to the mother of my boyhood pal I mentioned to you before, and I heard Stuart's mother say, 'I don't know *what's wrong* with John and Stuart! Stuart's in *trade* and John's gone *fish*ing!' " He was mimicking his friend's mother.

"What's wrong with trade and fishing?" I asked.

"Stuart did very well in business, but we were supposed to go into something *respectable*—either the Army or the Church," John replied. "The type I was raised among at Cowichan Bay had that British imperialist disdain for the sons and daughters of fishermen, farmers, and loggers and considered them 'dumb.' One reason I moved to Pender Harbour was to get among real basic people, although I don't know how much longer that will be true, with the bloody summer people, retirees from Vancouver, and tourists arriving and departing by plane in the harbor from morning to night in summer. I was homesick from the moment I set foot in England, and sick the whole time I was at school there. The school diet, what there was of it, was mostly starch—no thought of nourishment for growing boys. I grew two and a half feet in one year, and I had Bright's disease, mastoiditis—which left me with a sound like a waterfall in my left ear that I still have—scarlet fever, and pneumonia complicated by pleurisy. It just shows you what homesickness can do to the system. The school sent for my mother because I wasn't expected to live. The doctor in attendance had heard of a new operation perfected by a brilliant Canadian surgeon, Norman Bethune, in which a piece of rib was removed to drain the pleural spaces, and since I was expected to die anyway, the doctor tried it and saved my life. I revere the memory of Norman Bethune, although I never knew him. At the end of the term, I was supposed to come back to Canada and sit for exams for McGill University, in Montreal, but my illness ruined that. I might have gone there to study economics and history. I got to grade eleven and a half, but I lost the whole spring term and there were all those doctor's bills, so I couldn't take another term at Sedbergh and never got my certificate. My mother sent me to recuperate that spring at Eastbourne, on the south coast of England, with a friend of hers whose husband was a Fabian socialist. I was seventeen, and he made a deep impression on me."

"How so?"

"I became a ninety-nine-and-three-quarter-percent political animal," he answered. "What I hate more than anything are polite arseholes who agree; that's the road to destruction of mental protein. I believe in struggle—that physical and moral softness is *death*, and that we human beings can do *anything*."

He shifted in his chair. "I came back to Canada at seventeen, and spent three or four days in Toronto hunting for work, but it was the height of the Depression. So I returned to Vancouver and got a job as an office boy in the accounting department of the B.C. Electric Company at thirty-five dollars a month, and I was also feeding unemployed friends. We lived on the beach out by Siwash Rock, in Stanley Park. In three years, I advanced to seventy-five dollars a month, but the Depression was so bad they cut us back ten percent. In 1933, I think it was, the unemployed staged a march in Vancouver. They were marching peacefully through the center of town and the police were ordered to club and smash hundreds of the demonstrators. If I hadn't been there and seen it all, I might not get so bloody furious and be more easy politically, but I *cannot* forget. I *saw* the police smash those peaceful unemployed, many of them world-war veterans, and I ran and helped one or two of the wounded. That's an indelible memory in my mind. That picture lives with me. I have this chip smoldering below the surface, and though I know a big fault of mine is categorizing, that street scene is branded and burnt into my brain, and I am deeply class-conscious. I do put words in people's mouths, am not as objective as I should be, and am far too idealistic about humanity. Although if we *don't* believe in a deep human capacity for love more than hate we are *dead*, aren't we, dear?"

I nodded, afraid to break the silence. His shouted words still shook the air above us. Finally, I asked, "How much longer did you stay with B.C. Electric?"

"I quit in 1934, the year after my mother died," he said slowly. He looked down at his crossed hands in his lap. "I think I'd better tell you now. In 1933, she was living alone at Cowichan Bay. My grandmother died in the late twenties, and the relationship between my mother and my Aunt Dot deteriorated. They rarely saw one another, although I remained on good terms with both Dot and Tom always." He cleared his throat, started to rise, and sat back. "One day, my mother loaded her pockets with stones and walked into the sea from the beach below our house." He stood up. "They found her body a half mile away. Tom was the coroner."

He folded the chair and laid it in its place on the deck, gathered up the dishes, and turned to go in the pilothouse. He stopped. "I worry about all women who are alone and lonely," he said gruffly. He stood there lost in his thoughts. Then he added, "For myself, if I ever become incapacitated—and, with my bad heart, I have thought about that possibility—if I can so much as crawl, I will go back to the sea, where we come from, where I belong."

John went inside then, started the motor, and ran back to the dock at Namu, while I was washing the breakfast dishes. He tied up next to Jimmy Chambers again, and we set off on foot in search of Fred Welland, the electrician. We found him, a middle-aged sprite with red hair, bushy eyebrows, and glasses, in red-and-white checked pants and a green sleeveless down jacket over a brown shirt with the sleeves rolled up, sitting cross-legged on the roof of a troller and conversing with a group of fishermen on the dock. "I'm hearing what everyone thinks about my new sea-lion-scaring invention," Welland said to John by way of greeting. "It seems to be working pretty well. I'd like to show it to you." And then: "Did any of you see that American pleasure boat that was here last night? It was a huge yacht. You could hardly have missed it."

The fishermen had all noticed it, including John, and Welland said, "He's been waiting days here for me to come and repair his generator, and he finally got tired and left this morning. If it's a choice between a pleasure and a fishing boat, the smallest gillnetter comes before the biggest pleasure boat. The others can wait a day or two." He laughed.

Welland came down from the rooftop onto the dock, and went with us from the float along the boardwalk on shore. We climbed a short flight of wooden stairs and crossed a broad platform made of flat timbers along the whole side of a large white-frame warehouse. Wooden racks had been set on the platform in neat rows, over which several fishermen had stretched their huge cumbersome nets to mend them. Welland opened a white door in the wall of the building and we entered his workshop. He introduced me to his wife, Ida, a thin, light-haired woman with a gentle face, and while she and John made arrangements for him to bring in the ailing sounder, I sat down on an upturned box among radios and sounders that I recognized and a lot of other electronic gear that I didn't. Welland put a big black plastic box on the counter and said, "Here's my seal-and-sea-lion chaser." He turned to me. "Nets cost thousands of dollars, and the holes sea lions make can ruin a net, so people are always carrying dynamite, or guns to shoot them."

"I carry a gun myself, in case a sea lion gets tangled in my lines," John said.

"This device is really just an adaptation," Welland said, patting the box. "You have to find something sea lions don't like, or seals, either, and the one thing they don't like at all is a killer whale. I've seen them just get up and go when I've had this machine on; they don't stop to reason. It's an inbred fear."

"What's inside?" John asked.

"A straightforward tape recording of an actual killer whale, and you can adjust the volume from the boat," he replied. "Originally, we planned to develop an artificial call, but when we got voice recordings from the National Research Council we found that the whale call covered such a wide frequency range that it could not be simulated; it was too complex. Then we had to figure out a device, some kind of transducer to transfer the sound, and a box sturdy enough to stand fish-boat handling. It's ten by six inches, designed to float with just the top at the surface of the water and secured to the vessel by a safety line tied to the end of the boat around the end of the net. We first took it out into known sea-lion country off the south end of Calvert Island and put it on the boat of a fisherman whose net had been severely mauled. We got a message from him, by radiophone. 'It works! It works!' he shouted, and he got the machine free for trying it out. Just me and Ida and my son own shares in the company we've set up. The device can't be patented, but if we produce a good deterrent we'll get sales. A lot of people have tried, but no one has worked out a good design before."

John stood up to go and, looking around at all the radios, said, "God, I'm glad I can get rid of the noise these radios make by shutting them off, but I feel sorry for the packers, collectors, and Fisheries officers, who have *got* to have them on! There are some very deaf people in the fishing business by the time they are forty, with the Mickey Mouse, VHF, and regular radio, all three, and having to decide which one the voice they want to listen to is on."

"Noise, noise. All you fishermen complain about too much fishing noise," Welland said. "The other day, the fellow on the Fisheries patrol boat said, 'For God's sake, get the noise out; it's driving me crazy! I've got too much electrical equipment; put on more condensers.'

Well, I put condensers all over the place, and I don't think it was any better. In fact, I think I made it a little worse, but he thought it was better, and that was all that mattered."

We went back to our boat, and while John was talking to other fishermen I took a stroll on the float. The occupants of several vessels had brought out canning equipment and were putting up salmon in glass jars for the winter. Like bread making, home canning of salmon was a mystery that eluded me. I had the fleeting sense of inadequacy that I always felt whenever Canadians served me their home-baked bread. The first time I met John's son Dick I asked him if he thought I should learn to make bread. "No," he replied. "My father takes exceptional pride in his bread, and there certainly would not be enough room in your lives or that kitchen for *two* bread makers!"

In our home in Garden Bay, when John announced that he was going to make bread, counters were cleared and John's prized possession—which he referred to as "my old crank tub"—a large metal pail with a wooden handle that rotated a heavy iron kneading blade inside, was brought up from the basement. He would then start rounding up the essential ingredients, some of which varied with his mood. I once asked for his basic formula, and he said, "I mix a large can of Pacific Milk with two cans of warm water. Then I put in a handful of flax seed, two dollops of cooking oil, three dessert spoons of brown sugar, three handfuls of rolled rye or wheat, and three handfuls of wheat germ, in a tub, add one teaspoon of salt per loaf, figuring on four or five loaves, and add two level teaspoons of yeast that has been allowed to rise in a large mug half full of water with a teaspoon of sugar in it. Then I throw in teacupfuls of one-third whole wheat and two-thirds unbleached white flour, and knead the mix until it quits sticking to my fingers. The texture was never as good when I kneaded by hand as it is since I've been using my old crank tub. Why in hell this is, I do not know. I start my best bread at ten or eleven o'clock in the evening in the coldest room in the house, and leave it until between five and eight the next morning, when I reknead it and cut it into pans for the final rise. Sometimes I raise it in the daytime, but that spoils outside work. The longer rise overnight seems better. Why do you suppose that is?"

Early the next morning, he would turn on the oil stove, peek several times under the towels that covered the pans, until the bread had risen enough to suit him, and pop the pans onto the upper oven shelf of the old stove to bake, while we tiptoed around the house being careful not to slam any doors. When the bread was done, he would enthusiastically knock it out of the pans onto a counter, but he could never wait to sample it until it was cool. After several minutes of admiring the lined-up loaves and noisily sniffing the air with rhapsodic comments about the delicious aroma, he would cut two slices from a loaf, spread them with butter and honey, and carry the warm bread, two cups, and a pot of tea tucked under his arm out on the veranda in the sun, closing his eyes in rapturous anticipation of the first bite. His bread was substantial and firm, like its maker, and went with us everywhere: on the boat, on trips into town, to Vancouver Island, into the B.C. interior, and on hiking and ski expeditions up the mountains. He would cut a loaf in half and butter the end before he sliced it with the large bone pocketknife he always carried. I was told that in earlier days he made bread on his boat while fishing, but he never did that on the *MoreKelp* while I knew him.

I wandered down the float to the boardwalk, and passed a small building where the wife of a maintenance man had set up a laundry for fishermen. She told me there was a communal bathroom reserved for women from the fish boats across the way, and I ran back to the boat and got our laundry, bath soap, shampoo, towels, and a change of clothes. I dropped the laundry off with her, and walked over to the women's washroom to bathe. I was dismayed to see that the primitive shower stalls had no curtains. The cubicle I went into at the far end was not as dirty as the bathroom at Port Hardy, but it was not very clean, either. I spread paper towels over the floor of the shower, washed my hair, and was scrubbing my arms when I heard a man arguing with some women at the end of the room. He was intent on pushing his way in, and one of the women ran for help, but I didn't wait. I put on my clothes without bothering to dry myself, and fled back to our boat, where I shut the pilothouse door and finished my bath. John just shook his head when I told him what had happened, and suggested we pay a visit to his friends who managed the oil dock.

As a fire precaution, the oil dock was situated at some distance from the main complex of frame buildings. To get there, we walked up a hill on a raised boardwalk that followed the water's edge, past the manager's red-roofed white house and its impeccably kept garden, which had a large grassy lawn and a border of bright hollyhocks. Beyond, the boardwalk led through a cluster of small wooden company houses, sparsely occupied now by families of plant workers who came to Namu for the fishing season from the Indian community of Bella Bella, four hours away by boat. Over half the houses were empty, doors open, windows missing, in a fairly advanced state of decay, with garbage scattered below. We entered the woods beyond. Tall trees, chiefly coniferous, gave the cool woods a piny fragrance. Walkways extended for almost a mile through this deep forest, and the board paths, resting on wooden frames that raised them a foot or so above ground level, were so narrow that we had to walk single file. Large ravens in the trees called and answered as we traveled through their territory. The ravens' hoarse voices were interspersed with flutelike calls from other birds. Where the incline was particularly steep, there were white handrails to cling to. We had to watch our footing because of broken boards, and John said, "I'm afraid B.C. Packers is letting things fall into disrepair. One of the men told me that it is ten cents cheaper a pound to can in Steveston, at Vancouver, than at Namu. They have to bring people here, and maintain living quarters—a cook-house and so forth—whereas down there you can call in a crew who go back to their own houses and provide their own transportation."

How beautiful this was, how unlike anything I had ever seen! I kept looking down at the ground cover on both sides—at the delicate wild-flowers, the green mosses, the creamy velvet surfaces of the fungi growing right at the base of some of the trees—and had to run to catch up with John. We came out in the open, and the boardwalk broadened into a row of houses on a wide deck. All were boarded up but one, which the Japanese fishermen, who had their own dock away from the others, used as a private bathhouse and card room. We passed a sign reading "You Are Now Entering Sunnyville," and our plank path ended abruptly at a small dwelling of unpainted wood, with curtains at the windows. We stopped where the deck squared off, to look at the magnificent view of lavender hills and blue water, before we turned toward the house. Coming upon that tiny, neat structure at the very end of the wooden ramp was like being in the middle of a dream. "This is where Norman and Margaret Harrod live," John said. "Norman is in charge of the oil dock, which is just around the corner beyond. Margaret made those lovely earrings from seashells that I gave you for Christmas."

I had afternoon tea with Margaret, sitting in her minuscule living room, while John went on to the oil dock. She brought out a number of delicate shell earrings like the ones John had bought for me—tiny shells in various shapes and shades of white, tan, and pink, cleverly attached to earring fasteners. No two sets were alike, but she had managed to find two well-matched shells for each pair, and I bought several. Offering me slices of delicious lemon cake, which I devoured, Margaret told me, "My husband managed a station in the mining town of Trail, B.C., and then he got a job with B.C. Packers at Spider Island. I learned how to buy and ice fish there—very different from the life I led in my first marriage, when I lived in Toronto. I love my little house here, and hope we don't move again for a long time. When we came here seven years ago, I was picking up these lovely shells on the beach, and I made a set of them into earrings for the manager's wife. She said, 'Why don't you try to sell them?' So I did. Now I have a great little business."

Walking back from the Harrods', John and I arrived shortly at a small lake encircled by trees. A group of Indian youngsters were splashing in the water by a dock directly ahead of us, diving from it and scrambling back on it again, and two people were paddling a canoe on the opposite side of the lake. John walked along the beach and selected a place on the rough sand to spread out his best sweater, a gray knitted cardigan with pewter buttons, from Norway, which he had worn to call on Margaret Harrod. He lay down with his head on it and closed his eyes. I thought he had gone to sleep, when he opened them and said, "I become immobilized, paralyzed during a strike." He shut his eyes again, and this time he did sleep for a while. When he opened his eyes, he sat up, his arms around his knees, staring out at the lake. "In 1935, when I started fishing, there were literally dozens of fish canneries and herring salteries, plus several hundred small logging camps, scattered up and down the B.C. coast. There were from sixteen to twenty canneries at the mouth of the Skeena River alone, employing from forty to two hundred people each. Now there's only one, B.C. Packers' biggest northern plant, close to Prince Rupert. The federal government makes a religion out of efficiency through centralization. The whole policy of pushing people economically into large urban groups such as Vancouver, Victoria, and Prince Rupert is all wrong. It's a very dehumanizing process." He looked up at a raven flying overhead, cawing. "It would be far better for the country, for example, if Namu had a cannery with two or three hundred native and white workers every summer, with most of the money going back into areas where the fish are caught—namely, Bella Bella and Bella Coola, where the native producers have roots going back thousands of years. It's much more important to spread the money around in local communities here and keep them going than to have superefficient operations with people who have no roots on the coast and fly in and out by plane, buy apartments and houses in Vancouver, and take off for Mexico in the winter. The Indians I knew well were displaced from the fishing industry because they weren't equipped by nature to be as greedy as most white men, and it's only recently that they can get proper financing to have their own boats."

We got up to go. It was getting cooler, so John untied his sweater, which he had rolled around his waist, and put it on. "I love fishing as a way of life, although I am not as well satisfied as I used to be. All kinds of things of a semiconservationist nature that I'm not sure whether I think or dream go through my mind. A great deal about fishing depends on positive thinking and faith, believing that what you are doing to lures is correct and keeping your lures and hooks in tiptop condition. Fishing is very fussy, and I don't think it's possible to be too particular about hooks. I believe fish have a very developed sense of sight and smell. If you're in an area of the sea where there's a lot of strong acid from jellyfish or a lot of electrolysis, you get rust on your hooks. When I was younger, I used to change the hooks twice a day, and now it's about every three days, because I haven't as much energy. The modern stainless-steel hooks used with plastic hoochies and plastic plugs don't require changing for days, because there's very little electrolysis and they remain bright, but you still have to file their points every morning. However, if you are towing brass or bronze spoons, as I like to do, they would be ruined in a day by electrolysis without a nonstainless hook to work on. Without those blued nonstainless black hooks I use, my metal lures would be pitted from electrolysis in four or five hours."

XV

We went back to Boom Bay, and John spent the evening making repairs on the boat. At eight, he turned on the Mickey Mouse to hear the news about the strike relayed to the waiting fishermen on their CB radios by union officials traveling the coast on the UFAWU boat, the *Chiquita III*, and to take part in the discussion that followed. In the morning, John went to Fred Welland's shop and exchanged our VHF, which wasn't working right, for his repaired sounder. On the way, he said, "The oil needs changing, the anchor roller needs a weld, and now my phone seems to be on the bum. These bloody boats need endless upkeep, which is why I quit fishing fairly early in September. So many jobs on this boat—painting, for instance—must have good weather, and it's usually fine in late September and October, but it literally takes me all my spare time in winter to just bloody get out again by spring. Fishermen's lives are very bad for being possessed by possessions. We need highly complex boats; and then a home, with floats, sheds, lots of scrap, and new lumber, rope, metal 'junk,' and a truck. I'm not complaining, because I'll bet if I didn't have my headquarters, my anchor of a home I love—with the idea of permanence it represents—you might not find me so irresistible!"

"I'm afraid if all you had was a leaky rowboat, I'd still find you irresistible," I replied.

While John was installing the sounder, I baked orange bread, and then I went to the store and bought some pork chops. Walking back and forth, I listened to the hum of voices coming from the boats lining the floats. There were a few pleasure yachts mingled in with the fishing fleet; one out of Seattle, Washington, flying the American Stars and Stripes, had picture windows through which I could see what appeared to be a whole living room, complete with sofas, easy chairs, tables, lamps—and a large, well-stocked bar.

I walked back to our boat, to find John talking to Mary Rossiter, on *The Venture* across the way, about what colors to paint their respective boats. "Bright yellow," I suggested immediately, and John said, "It'll be whatever color is easiest on my eyes, which is green."

The Gardner was running when I climbed over into our boat. We moved away from the *Kitty D.*, and I assumed that we were going back to Boom Bay, as it was almost suppertime. Instead, we headed out into the channel, away from Namu. We ran along for several minutes in silence before John spoke. He said, "There's an Indian fish cooperative at Bella Bella, and the members are still fishing. We're going there now. Is that all right with you?"

I had never been consulted before about where we were going. I liked not knowing; it was all new, anyhow. "Why do you ask?" I said.

"We're going to do picket duty outside the Indian Fish Co-op," he replied, looking at me as if he were searching my face to see what I really thought. "I wasn't quite sure it would be fair to you, but I didn't feel I could say no when the strike committee asked me to go. Well, I really volunteered, because we've only got one man there, Cliff Gissing, for whom I have the greatest admiration, and I want to be with him. Although our union may not be the answer in the present economic system in North America, the most destructive thing working people can do is to take a course of action that splits us from one another, and the whole labor history here, as I see it—particularly since our union became established—is for the monopoly that controls the fishing industry discreetly to divide us. I respect the Indians, and if in turn they have any respect for me, I want to try, by my presence at Bella Bella, to keep some line of calm negotiation open. I also want to make sure that there are no irresponsible actions or drinking of alcohol by anybody picketing to make a sad situation worse. If you mind going, I can take you back, but I thought maybe you'd want to stay with me."

"I can't imagine doing anything else," I said. "I don't understand very much about what's going on, but I want to stay."

"You're not afraid?" he asked.

"No, I'm not," I said. "I'm only afraid of what you'd say if I opened another milk can upside down."

It took four hours to run to Bella Bella. John was silent for a long time after we left Namu. I made tea and brought it to him, with digestive biscuits, which I liked almost as well as Bourbon Creme cookies, and John stood up, so that I could get past and sit down beside him. After he had finished his tea, he said, "My overriding feeling is that our union could have done far more in educating the rank and file of membership in the Native Brotherhood to the need for unity in the fishing industry. I also feel that, by comparison with any other union in Canada, we've had an excellent record on fighting discrimination on jobs in commerce and housing for natives—especially our leadership in getting a welfare fund that encompasses health, shipwrecks, and sickness. I personally have many good friends among the native people, who have taught me many useful ways of being more observant and a better fisherman. Through knowing them and watching their actions as fishermen, I have learned a little of that necessary factor to think like a fish and I feel that in turn I have never taken advantage of them throughout my fishing life. I have been as mixed up as they were, but I have tried to show my respect by attempting to help them to help themselves. I worry about whether the action I am taking through the union now of picketing will risk that friendship. I'm willing to take that risk."

The sun went down while we were traveling. It was a soft, blue evening, and I went outside several times to look at the running lights that reflected green and red shadows along the white walls of the pilothouse, and to glance up at the bright twinkle of our mast light. The water glistened black, and the white foam rushed up to the sides of the boat. We were passing mainland and islands that were impossible to separate from one another in the oncoming darkness. John pointed out the Indian village of Bella Bella across the channel, and the store and oil wharf run by B.C. Packers. We rounded a bend, and he slowed down and, as the Indian Co-op's cannery—low, rectangular buildings and hangars that had been part of an Air Force base in World War II—came into view ahead of us, said, "We're here. This is Shearwater." He stopped beside a lone troller anchored in the middle of the channel, with the name *Shaula* in black letters on the hull—a big white boat with lavender exhaust pipes, and large signs facing both ways tacked to its mast that read:

SALMON FISHERMEN
ON STRIKE
United Fishermen and Allied Workers Union

A smaller sign along the side of the boat said:

Picket
U.F.A.W.U.

"It's lucky Cliff delivered all his fish at Namu, or we wouldn't be able to tie up to him," John said. "He's got a refrigeration plant on board, and they make so much bloody noise, nobody can sleep. More than once, I've had to move when I found myself anchored anywhere near one. I don't know how Cliff stands it."

While John was tying the *MoreKelp* up to the *Shaula*, Cliff Gissing came out of the pilothouse, rubbing his eyes. A lean, youthful man in his fifties, with clipped brown hair turning gray, wearing a red-and-green plaid shirt and beige whipcord pants tucked into rubber boots, he said, "I only stopped patrolling back and forth a short while ago, and was taking a nap until you came. One boat went past shouting obscenities, and several people have been out to see me, but otherwise I've been left pretty much alone." His English accent was so broad that I had trouble understanding him, and he told me that it was East Anglian; he came from the county of Suffolk, on the North Sea.

Cliff went below and got two strike placards similar to his, which John tacked to our boom. We watched fishing boats go in and out of the lit-up cannery complex. A third troller, the *Duby C.*, joined us; its owner had his twelve-year-old son with him. At ten, John and the skipper of the *Duby C.* went over to Cliff's boat to talk over the radiophone to union officials. We spent the night tied together, with the *Shaula* in the middle. John and I were so used to being alone that all sounds seemed strange. The boats creaked and groaned against one another, and we felt compelled to talk in whispers. It was eleven o'clock when I climbed into my bunk for the night.

At seven-thirty the next morning, while we were eating breakfast, John noticed a troller heading in our direction. He picked up the binoculars to get a closer look and said, "That man's a former machinist at Namu. He's got his wife and two sons with him, and it looks as if he's coming to see us. There are two other people with him." He took a second look through the glasses and said, "Why, that's Bill Bailey and his wife, Daphne. Bill and Daff were trollers themselves until two years ago, and now they are managing the store and oil dock here at Bella Bella for B.C. Packers. Daff has quite a story: she grew up on Calvert Island and her family, the Grextons, were the only people living there. One Christmas Eve, her father and brother disappeared in a terrible storm and were never found. Daff, her sister, Doreen Kaisla, Kal's wife, and their mother were stranded there for weeks in midwinter before rescue came."

The boat, the *Menzies Bay*, drew up alongside, and the skipper hung on to a rope John threw him until the Baileys climbed aboard. He departed, saying that he would come back for them in an hour or so. Bill Bailey was a quiet, stocky man with glasses, and Daphne, a large, hearty, good-looking woman with upswept dark hair and a warm, friendly manner. "We wanted to see what we could do for you, and to say hello," Bill said.

Cliff Gissing and the fisherman from the *Duby C.* climbed over from their boats to talk about the strike with Bill, and they all went into the pilothouse. I brought out the teapot and orange bread, and Daphne Bailey settled down in the beach chair on deck and I on the fish hatch. She was curious to know how I liked living on a fish boat. "We've known John a long time," she said. "We were really surprised to hear you were on the boat with him. I guess everyone was."

We both laughed. I said, "I love living on the boat, and I've heard about you, too. Were you really brought up on Calvert Island with no other people around but your own family?"

Daphne replied, "My father, mother, brother, sister Doreen, and I lived at Safety Cove, on Calvert Island, until I was sixteen. Just the five of us, on the whole island. Before the First World War, a group of German settlers were there, but they left for the United States when war was declared, to avoid conscription or being deported. So we were the only ones. I never went to school and it was terribly lonely. My father had various ideas: starting a cannery, raising crops, cattle. Nothing really worked too well, but Father did have traps. I learned to read, but I have no memory of being taught to read or to write. We stayed on Calvert Island until my father and brother died. They had gone fishing in my father's gillnetter and were supposed to be back that evening. My brother was eighteen then—I am fifty-three now, so that was thirty-seven years ago—and they were down by Grief Bay, at the south end of the island, checking their traplines or fishing. Something happened, because they never came back. We waited twelve days, without knowing, wondering and watching for them, and then I tried to row across the Sound for help. It was winter and the weather was awful, so I had to come back. Later, I saw a big seiner, and I rowed out to it in a terrible wind, I don't know how, and when I arrived someone said, 'God, it's a girl.' After that, everyone looked for my father and brother. They never found a trace of them, except for parts of the boat, and there were some remains of a fire on the beach. I think my father drowned. He always got arthritis the minute he hit the water." She paused. "I think my brother either got caught in quicksand or was eaten by wolves or died of exposure." There was another pause. "Of course, my mother and sister and I couldn't stay there alone, so we went to live in Bella Bella, and I've been there ever since. My mother died about five years ago, and Doreen and Kal moved to Delta, near Vancouver, when their children were in school."

"Do you think living like that had any special effect on you?" I inquired.

She looked grim. "Oh, yes, it was *really* terrible," she answered. "I like people, but I still can't stand too many of them at one time, and when I am in a big crowd I have to go off by myself and sit down."

When the men came out of the pilothouse, John was carrying the saucepan full of hot coffee. He poured some for everyone, and while we were chatting he said to Daphne, "A couple of years ago, I went ashore on Egg Island and stumbled over a grave. Do you know whose it was?"

"Oh, yes. That was the lighthouse keeper," Daphne said. "He had an argument with his wife, and she left. After a month, she was very lonesome for him and sent him a telegram that came through the Bull Harbour radio that said, 'I'm coming back.' Somehow the message got garbled en route, and when he got it, there was some misplaced punctuation, which made it read as if she was not coming back. He committed suicide the day before she returned. She thought she wanted to stay on as lighthouse keeper, but she only remained until the government got a replacement."

After the Baileys left, we continued to sit for a while on deck. Four boats arrived in rapid succession, tying up briefly to talk. After lunch, a motorboat came by very fast and made an enormous wash. There were three Indian youths in it, sitting up very straight, with their backs to us, while we were swaying heavily from side to side in the waves their boat had made. For the rest of the afternoon, we were left alone, except for the arrival of a small troller with the green-and-orange B.C. Packers flag tied to its mast. At suppertime, another troller joined us on the picket line—the *Five Spot II*, a beautiful modern vessel, whose conveniences included a shower bath and comfortable-looking chairs. The owner, Al Schaible, told us, "This is my fifth boat but my second *Five Spot*. I got the fiberglass hull and built all the rest myself."

XVI

It was hot out the next morning, so we spent it sunning ourselves on deck. Fishermen, friends, and people who were just curious came out to see us, and Indians scooted around us in outboards all day. A ferryboat from Bella Bella ran up close enough to the *MoreKelp* for passengers to take pictures of John and the boat. I made a fish stew for all three boats for lunch, and baked an apple coffee cake from my *Five Roses Cookbook*. Cooking made me feel useful.

While John was taking his nap, lying as usual at full length on his back on the deck with a wool sock over his eyes, Cliff took me on a tour of the *Shaula*. I complimented him on the spaciousness of its pilothouse, the racy effect of its yellow linoleum, and his lavender exhaust pipes. We sat and talked on his deck, so as not to disturb John, and he told me he was from the English coastal town of Aldeburgh. "My father farmed and fished, but he was a very poor fisherman. I've spent all my life at sea—since I was ten and a half," he said. "I've never even worked on land, but I've always felt closer to the land than the sea, even after a long trip. I wanted to be an airline pilot, but my education stopped at grade eight, so I joined the English Navy in World War II and was sent to Malaya. I once went five and a half months without landing. The longest I've been on this boat is twenty-six days, which is too much of something. You have to have control at sea, but you can relax on land."

He had been rubbing his arm while we talked. "Did you hurt yourself?" I asked.

"I did," he replied. "My arm is always sore from the time I fell into the hold and fractured my back. It hurts all the time."

He was silent for a minute or two, then continued, "I came to

Canada twenty years ago. I had been fishing cod and herring in England. When I was in the Navy, I used to go beam trawling on my holidays. We sailed after the pub closed at ten-thirty, and dropped anchor, ready for fishing, at seven Monday morning. Fishermen in the North Sea were more relaxed in their attitude toward harvesting the resource than our fishermen are here. The resource in the North Sea would never have held up if they had harvested their fish with the same intensity and pressure. As I see it, the whole ocean is a fishery, and continental-shelf fisheries of individual nations have to come under world management. I feel that the fish and mammal protein resource of the oceans belongs to the people of the world, not to the country that has the ability to harvest the resources of those oceans. I see no reason why the harvesting shouldn't be carried out under the jurisdiction of a world body, and part of the catches distributed through the United Nations to impoverished countries without the ability or means to carry out a fishery themselves. Countries should draw fishery closing lines to protect their domestic stock only by mutual agreement. The remainder of the oceans and continental shelves should come under the sole jurisdiction of a United Nations world fishery body, to keep the resource at a sustained yield for the future."

Cliff was now leaning forward, talking rapidly. "For example," he continued, "one can fly along the east edge of the Grand Bank of Nova Scotia right down to George's Bank along the Atlantic Coast for a thousand miles at night and see complete cities of light—a series of cities. They show quite conclusively the tremendous pressure being put on the fish resource, largely by oceangoing fishing and processing vessels. Once you have seen and understood the capability of these huge fleets, the end result of failing catches due to overfishing in the North Atlantic is fairly obvious. This spells out quite simply the need for a world convention on fishing matters."

John had awakened and joined us, listening quietly. "The first fifteen years I fished, most of the time I didn't know where I was, but with electronics you can really find the fish," he said. "You can know too much about the poor wretched fish and everything else in this world. I personally think we are working on the last of the wild stocks."

Cliff frowned, nodding agreement. "As for management of the resource," he said, "I see no hope of any change with past policies as long as the responsible people give way to political pressure groups—in particular, the large commercialized sportfishery. A conservative estimate would be 275,000 actual sportfishermen a year in Canadian waters, and they have a large political boot. Because of them, most government hatcheries are located within a reasonable distance from a city. Being so close to cities and industries means that most hatcheries are in the most polluted waters of B.C. Those areas are also vulnerable to heavy American sport and commercial fisheries, as well as the heavy fishery by sport people here. At the same time, under the guise of conservation, commercial fishermen are severely regulated. What should have been done and should be done in the future is to spend salmonid-enhancement monies in more strategic areas on the B.C. coast, where there is less pollution and the fish are not so vulnerable to interception by U.S. sport and commercial fishermen."

We were deep in conversation when a white boat came up beside us. It was the union's boat, *Chiquita III*. A man on the deck threw a line over. John and Cliff greeted him warmly, climbed across, and disappeared into the *Chiquita*'s cabin. I retreated to our pilothouse with the idea of working. I sat down for a moment on John's bunk, and when I moved away the pillow fell forward, and with it a nine-by-twelve pad of paper behind it. By now, I was quite accustomed to finding things stored in unexpected places—especially under and around John's bunk—an area which we both now referred to variously as his "office" and "the library." *The Complete Works of William Shakespeare*, wrapped in newspaper, was always to be found under the mattress corner at the head of John's bunk. I also had an "office" of my own; the space under my own bunk, between the wooden frame and the mattress, was ideal for storing papers that had to be kept flat and dry. I was about to put the pad back behind the pillow when several sheets of paper stapled together fell out. The pages were curled around the edges and brown with age. A FISHERMAN'S MEDITATIONS was the title on what appeared to be a Xerox copy of a report, and in the right-hand upper corner, in John's handwriting, were the words: "For Fishing Course 10, 1961–62, Pender Harbour High School."

I settled right back on the pillow and proceeded to read. The report was loosely divided into several subjects. The first subject was *Mental Approach*. Under that heading, unmistakably in John's style, I read: "It is a great mistake to ever think there is some big secret or final solution to catching fish—there isn't. The successful are constantly trying different hook weights, black and stainless hooks, different leader lengths, different depths and speeds, spoons, plugs, flashers with bait or flashers with herring . . . Always remember that what catches a fish today may repel him tomorrow. I always remember the man who painted an egg-wobbler spoon white enamel, put it on his gear among thirty or forty plain brass and red egg wobblers. It caught twelve or fourteen big Cohos within two hours. He stayed up losing precious sleep to paint twenty egg wobblers white, baked them to set the paint, put them all on the gear at 4 a.m. the next day, and at noon took them off in disgust. They had caught no fish, but the old reliable brass were fishing steadily! He told me, 'Don't get the idea you've ever solved the problem of catching fish; let them tell you and keep an open mind!' "

I got up and made myself a pot of tea, rummaged around until I found an unopened box of Bourbon Creme chocolate cookies that John had hidden in the back of the cupboard, took off the bunk canvas covering, pulled John's red blanket over me, and continued reading. "I lost a good fifteen hundred to two thousand dollars over two or three years before I got it through my thick skull that *fixed ideas* about movements of fish were wrong ideas," John went on. "You must learn to weigh *all* factors of weather, water temperatures, 'feed,' such as herring or needlefish, tide, size of fleet, and fishing effort. There is a factor some men develop and others have naturally and still others never seem to attain—that elusive and indefinable instinct, plus the ability to be able to think like a fish. I believe those humans who attain this somehow reach back into time and their evolution and 'tune themselves in' on those instincts we once possessed for survival several millions of years ago."

Balancing my mug in my lap after taking a swallow of tea and a bite of cookie, I read on: "As your experience increases, your brain will know what to retain and reject. Remember that what works for one man and one boat may not work for another. Don't allow yourself to 'give up.' "

Then he warned: "If you do not like (or enjoy, as I do) being alone but must be with crowds or need much company, do not be a fisherman. You will be most unhappy, as, due to wind and fog and the very nature of your life, you will be alone a lot. I have kept my fishing days and hours since 1952. My total hours traveling to and from grounds, fishing, plus overhauling boat in winter, add up, most years, from fifty to one hundred hours more than the man who works at a full-time job . . . Do not fall prey to the disease of always having to beat the other fellow or be high boat, because this road leads to unhappiness and discontent . . . Keep trying to improve your methods and be happily convinced that you will always do a little better, if not in fish and dollars, then in greater experience and knowledge."

Next, *Gear Stocks*: "I advise a wide variety of gear be kept aboard, but to build this stock is extremely expensive, so build slowly. Time and again, I have seen fish bite a certain spoon or hoochy or flasher or plug and the man that has them catches the fish. I was fishing beside two boats one day. I had sixty coho, one had one hundred and fifteen, and the other, one hundred and thirty-five. I had neglected to stock up on a certain type of 'hoochy' lure, and they had done so. The closest supply was four hundred miles away (in Vancouver). This went on for several days . . . I check all my gear stocks every winter, and do my 'shopping' for gear then, not at the last minute, when there are often shortages."

About *Lures*: "If plugs work, they are better for springs later in the year than April or May. I also believe the fish bite them in the afternoon and spoons in the morning, but this is just conjecture . . . Developing 'an eye' for the curve of a spoon, just as an artist or boat builder develops 'an eye' for the perfect picture or curve of a hull, is a most useful attribute. I have seen trollers who developed a skill with one particular brand of spoon which was almost uncanny, through years of keen observation and *great faith* in their particular lure. That faith is most important." He suggested polishing spoons occasionally with steel wool or a silver-plating compound instead of Brasso, and trying different hooks and speeds. "One man cut the swivels off his spoons and it helped . . . for a dull day. Another man increased his production one day by attaching a tiny bit of red Scotch tape to his spoon. Experiment!"

Under the heading *Boat*, he stated: "*Quietness* in the hull machinery is *most* important, and, if building a new boat, design it so that there is a lot of water around the propeller, which must be true and balanced and not whistle, and the rudder must not rattle." He favored "depth and narrowness of beam, rather than shallowness and width" (like the *MoreKelp*, I thought). "The deep narrow ones roll far slower, the gear works far better, and one has far more comfort whilst trolling, but usually they are much poorer boats in a following sea." He preferred "a good boatbuilder who also fishes himself," rather than a marine architect. "Keep your ears and eyes open and look at many boats, because any mistakes you make you must live with for years. I know—I've had five boats. When buying engines or gurdies or sounders, *don't listen to salesmen*—talk to those who've *used* them, and *machinists* and *mechanics* who service them. *They know.*"

For *Hours and Time of Biting,* he quoted a Department of Fisheries study on why some fishermen get more fish: "The main conclusion was that the man who had his hooks in the water the longest number of hours was the one who got the most fish. Trolling daylight to dark is your best chance."

At the end, he wrote: "Don't ever forget that you are an 'above the water' animal, attempting to outwit a salmon who has brains or instinct so fantastically developed that he can swim south from years feeding off Alaska, and home in on his own river in British Columbia or Washington. His life and habits are of the ocean, of which we are learning a little but have much to learn."

The door slid open behind me, and John asked, "Where were you? Didn't you hear me call?"

I shook my head, holding up the sheets I was reading.

"The strike committee has given us permission to travel," he said. "As soon as someone arrives to relieve us, we'll go in to Bella Bella for supplies at the store, and we'll be on our way." He looked at the half-empty box of cookies. "Oh-oh," he said, and grinned, wagging a finger at me. Then: "You found my paper for the fishing course we gave at the high school in Pender Harbour!"

He poured himself a cup of tea, reached into the box for a cookie, and sat down on the end of the bunk. "A fishing course!" I said. "Tell me about it."

"I was on the school board in Pender Harbour, and in the late fifties and early sixties we had a lot of educational problems, the worst being at our high school, where we had two academically well-degreed principals whose understanding of students was nil," John began.

"We then got wonderful Frances Fleming as principal, who as a teacher had been holding the high school together to keep it from collapsing for several previous years. She had what I consider great educational leadership. It's much easier to succeed with a student whose home conditions are positive, but we had many students whose home background was uninspiring, to say the least. Under her positive self-confidence-building influence, we noticed that they blossomed into useful members of society."

He gulped his tea and continued, "At that time, we had no shop for the boys, so she got together about forty of us fishermen in the harbor, and tried to see if we could set up a course on fishing in place of a shop course. She took everything we said, organized it into a course, and sent it to Victoria to get it approved for credit. It was the first locally developed course in British Columbia to get provincial approval. Up to then, the provincial curriculum was rigid, developed entirely in Victoria, but in 1960 the government said if any school could develop a good enough course they would give the students the credit toward graduation. Ours was the first. Sean was still in high school then, and I was a very active member of the Parent–Teachers Association."

He leaned over and took another cookie. "The fishing course was entirely her idea, but she consulted me a lot," he went on. "We had a meeting about it with other parents and I got up and talked. I said, 'In making this kind of a decision about your child, it's your intention, your goodwill as parents, that counts. As long as you try to make the best decision you can, don't regret it.' I remember saying, too, that in our family, when money comes in, the boat has to have first claim on it, and that my children accepted this because they understood its importance. Mrs. Fleming felt that the boys should have, as she put it, the 'mental set' necessary for a fisherman, and she asked me to teach that. First, I wrote this paper, and I presented it at a meeting when Mrs. Fleming tried out the course structure on us fishermen. She wanted to see if she was on the right track before she typed out the course and sent it to Victoria for credit. We sold Victoria on the idea, and the next September we started the course. It was called 'Fishing,' and we broke it up into units: rules of the sea, elementary navigation, galley cooking—the school used a little kitchenette that was off the gymnasium, showed the boys how to preserve food and get a meal, and instructed them on their nutritional needs. One unit was rope and cable splicing, and net mending. They went to one of the lofts in the harbor to learn that. Another was first aid. And there was diesel maintenance, troubleshooting, and emergency repairs. We brought in experts from our harbor and Vancouver. I got a mechanic to come up from Vancouver. It was a two-year course, and we only touched the surface, but the boys knew the basics of fishing when they were through. Girls just didn't do that sort of thing then, so there were only boys, aged fifteen to seventeen, in the class. It amazed us how adolescents responded, in the context of past familiarity with the subject. Mrs. Fleming had *A Fisherman's Meditations* copied and gave it to the youngsters as part of the course, which only lasted about four years, sad to say, until Mrs. Fleming left to become the first woman superintendent of schools in British Columbia and, later, a member of the Ministry of Education. The school put in a shop with traditional teachers where the kids made traditional mailboxes, and so on, and the course faded. I really regretted that. The boys who took that course were most valuable as assistants to fishermen."

John stood up as we heard a boat approaching. He stuck his head out the door and said, "The relief picket boat is here. I'll go over and talk to our relief man and then we can go. Cliff is ready to take off now." He sighed. "God, how I wish this bloody strike was over! We'd all like to get back fishing."

I straightened John's red blanket and pulled the canvas up on the bunk, put his *A Fisherman's Meditations* back where I had found it under the pillow, washed our cups, and hung them up. By the time I got out on deck, John was untying the *Shaula*'s lines. We said goodbye to Cliff, who was going back to Victoria until the strike ended, and to Al Schaible on the *Five Spot*, who was staying for another day or two of picketing.

XVII

After a short run—it seemed like a matter of a few minutes—we arrived at Bella Bella, where we filled our fuel and water tanks, mailed our letters, and bought a few groceries. We also washed our clothes and ourselves in Daphne Bailey's sparkling-clean washhouse, reserved for commercial fishermen and their families, and bought a custard pie and several loaves of bread she had just made.

John glanced up at the cloudless sky as we set out again. "A perfect day. No rough seas to wear me out, which is just the way I love it," he said. "We're going to a town called Ocean Falls, a couple of hours away, where it rains all the time. The author of a book about it called *The Rain People* calculated its annual rainfall—it's a *really wet* place—at a hundred and seventy-two inches, or fourteen and a half feet."

It was late afternoon when we turned up a narrow waterway—Cousins Inlet on the chart—and, coming around a bend to its extreme head, we arrived at Ocean Falls. An immense glittering curtain of water thundered down to make a sea of swirling white foam at the base of the falls before it swept out into Cousins Inlet. "Normally, the water goes through turbines in the powerhouse to make power, but when they've got too much water they open the sluice gates and spill the excess over the dam. We got here at the right time, just when they were sending the extra water over the top of the dam," John said. "I was hoping you'd be able to see this when we came in. It's a spectacular sight."

The roaring waterfall was the divider between the residential and industrial sectors of this phenomenal "town" crouching at the base of the high steep hills on either side of the dam. While we were still approaching the harbor, and, with the help of the binoculars, John pointed out the terraced rows of houses and apartment buildings, some from four to eight stories high, on the precipitous slope on the left, and the ascending plank roads laid out between them, three or four times the width of the wooden walks at Namu. I asked about a prominent six-story cream-colored building at sea level, and John said that was the hotel, and that the flat space it was on also held a store, a hospital, a post office, and a customhouse. On the right of the waterfall were low buildings containing the *raison d'etre* for the existence of Ocean Falls: the pulp mill, with its attendant warehouses, sorting shed, sawmill, powerhouse, oil tank, docks, and cranes.

We tied up at the government float below the residential area, where we were surrounded by pleasure boats that John said belonged mostly to Ocean Falls residents. We had passed several of them fishing or sailing around in the area as we approached. We stepped out on the dock. The earth shook and the air rumbled with the impact of the water going over the dam. We walked up a plank street, which I could see now was built on stilts, until we came to the hospital, where we stopped at the trim outpatient clinic while a young African doctor, who had come to Canada for his medical training and was serving his residency at Ocean Falls, examined a rough spot on John's cheek. Skin cancer is one of the hazards of fishing because of long daily exposure to sun. The doctor reassured John that all was well, and advised him to grow a heavier beard to protect his skin. We stepped outside again into a light drizzle, although the sun still shone. For the rest of our walk, up the wooden streets, past the wooden houses and apartments and higher up still, past a small theatre and water tower to the dam itself, the rain continued to fall softly.

As we were nearing the top of the steep incline, the rumble ceased. Silence came fast and unexpectedly. "They've closed the gates," John said. "Come along. We can see the dam now, and Link Lake behind it." He took my hand, and we climbed silently up to the top and a bit beyond, to get a view of the long and narrow body of water—Link Lake. Walking back, we stopped to look at the dam, exposed without its shimmering curtain of water as a massive, ugly wall of rough concrete. From the top, we looked down on the roofs and stacks of Ocean Falls, founded in the first decade of the twentieth century, because of its easy access to waterpower and admirable conditions for a mill site.

John picked some rosy-pink berries off bushes along the path and gave them to me. "Try these," he said. "They're salmonberries. The

Indians make a good dessert out of them—a kind of pudding." The salmonberries looked like raspberries except they were a bit bigger, paler, and, to me, they tasted watery and flat. We continued to pick and eat them as we ambled down the planked road, standing aside when an official truck rattled past.

The first object to catch my eye at the entrance to the hotel was a telephone booth. I am always shocked when I see a telephone booth in a wholly remote place like Ocean Falls. John wanted to order parts from Vancouver sent to Namu by air for the boat, and I wanted to call my sister in Cleveland. There was such a long line at the booth we decided to go into the Chinese restaurant there and have supper instead. The restaurant, an unadorned room whose chief attraction was its view of the waterfront, was filled with patrons. It was run by a Chinese family, old friends of John's, who gave us an enthusiastic welcome and an excellent dinner. The mother, a son, and a daughter took turns sitting with us when they were free, and urged us to come to their house after the restaurant closed at eight. We stopped in briefly at the house of the restaurant owners a little later. We walked up the plank road to a simple suburban-type bungalow and had a quick cup of tea with them. Their kind of hospitality in an isolated place like Ocean Falls meant a lot to John, who had fished alone so long, and the warmth of their welcome touched me, too. When we left them, the rain had turned into a downpour, and I opened a small red folding umbrella over my head that I had brought from New York without a thought for the difference between my being five feet two inches tall and John's height of six feet four inches, until John said, "You are a menace! I'm going to lose an eye if you don't put that thing down or let me hold it over both of us." He carried it so high that the rain came in underneath, so he folded it up and tucked it under his arm. It was pleasurable to walk "home" to the *MoreKelp* hand in hand with John.

XVIII

We awoke in the morning to rain brushing against the windows. Right after breakfast, John started the Gardner and we ran back down Cousins Inlet to a marine crossroads; this time, we took the left-hand waterway, the much wider Dean Channel. Our goal, three hours away, was a pocket-sized opening called Eucott Bay. John had obtained written permission from the strike committee before we left Namu to food-fish for ourselves while we were traveling. It was almost noon, and I was preparing a salmon chowder with one of two small cohos John had caught, along with one fair-sized spring salmon, fishing briefly on one line on the way to Ocean Falls. He looked in the red stewpot on the stove. "We'll take some of this to two old fellows who live in a float house at Eucott," he said.

Almost immediately afterwards, he slowed down and turned left into a small bay hiding among the trees and stopped a few feet inside. At the far end, I could see what appeared to be marsh grass and a sandy beach. Beyond, through a V in the low hills covered with green firs, I glimpsed snowy mountains blending with a mass of white clouds floating across a bright-blue summer sky. It had stopped raining.

John anchored, and pulled down a chart. We were in fourteen fathoms, but directly in front of us it was only three. He said, "We'll have to wait for high tide to call on Frenchy and Simp*son*, because right now it's low tide and their dock is resting in mud." He pronounced the latter name with emphasis on the *son* to give it a French twist. "They live in a float house, an old abandoned fish camp that's been pulled up on the grass. It's just out of our sight now, beyond those three pilings over there." Farther ahead of us, perhaps two hundred yards away, where the chart showed a depth of half a fathom, was a small unpainted frame house on floats with a ramp to the beach. "A retired fisherman lives there, when he's around," John said, "but he doesn't like visitors much. Don't mention him to the old men when we visit them. They aren't speaking."

"Who else lives in this bay?" I asked.

"That's all," John said. "All winter long, just the three of them are here, except for an occasional visitor from Ocean Falls or when the United Church boat, the *Thomas Crosby*, stops in to check up on them."

"There are only the three of them in this bay and this man and the two others don't speak?"

"That's right."

"Why?"

John shrugged. "It could be anything, but I think it was a political disagreement. It doesn't matter; they may not even know themselves. These guys living in isolated spots like this tend to get paranoid."

In the afternoon, John put the dinghy over the side and said, "It's high tide, so we can go over to see Frenchy and Simpson." He poured some of the salmon chowder into a container, and put it in the dinghy along with a bag of apples and oranges, the custard pie, a six-pack of beer, and the bottle of Scotch he had bought at Ocean Falls. At the last minute, he went below and got the spring he had caught on the way, and brought that along, too.

I took off my Stanfield's, put on my red turtleneck sweater, and got into the dinghy with John, who was resplendent in his Norwegian cardigan. He started rowing us across the bay toward the three pilings. Halfway, John stopped rowing and dipped an oar quietly in the water to turn us around so that we could both see the far shore. "Don't move. If you look in the water toward the V in the trees, there is a seal making eyes at you," he whispered.

I saw a dark-gray triangle in the water. The point at the bottom was a nose, and in the wide part two eyes were looking steadily at us. As noiselessly as the seal's head had appeared, it was gone, resurfacing at some distance on the other side of our boat. When John turned us around, it disappeared again, reappearing where it had been before. The seal was playing games with us. Next, it appeared beside the *MoreKelp*, which was swinging around slowly at anchor. John quietly turned us around, rowing toward the boat, and the seal vanished. We sat and waited to see if it would show up again. It didn't.

"The seal population is unbelievable since shooting them has been forbidden," John said. "Seals eat colossal amounts of herring in every little bay, and a lot of our salmon. They need to be shot off again." When I protested, he shook his head. "The balance of nature's been buggered. We *have* to bump off seals to survive, but I don't feel good about it. During one strike in the nineteen-fifties, I tied up in Winter Harbour, on the west coast of Vancouver Island. I gave my deck two or three coats of pine tar and linseed oil—I think paint rots wood—and when I'd got the decks done I went into Quatsino Inlet and stayed with a friend who had a fantastic garden, at the height of the garden season. He had strawberries, raspberries, and green beans, and he produced seven to eight thousand pounds of holly off his place every year because he used seaweed as his fertilizer. The water was so warm that year that we went swimming and played with a tame seal that belonged to the store. This seal made a daily visit to all the floats. Everyone kept something for it to play with, and I had a piece of trolling pole that I would chuck in off the float when I saw the seal coming. It would grab it with its flippers and go round and round. Some s.o.b. didn't know about our seal and shot it. It used to come home every evening to the store and crawl up and lie down behind the stove, and this night it came home wounded and died."

John started rowing again across the bay toward the three pilings. There we entered a channel in the flat, grassy marsh. At the far end, half hidden among the trees, was a low, flat-roofed, oblong wooden building with a covered porch. It was dark red, with white trim on the eaves, the frames, and the moldings around the six panes of glass in each of its small square windows. The building, a foot or so off the ground, was on a scow that was on skids; and in the grass behind were an indeterminate assortment of half-ruined, unpainted small shacks. We came slowly to the rough plank dock in front of the scow as a slight, somewhat bent little man rounded a corner of the porch, hurrying toward us. He was wearing the usual dark wool fish pants, held up with suspenders; and a yellow, red, and black plaid wool shirt wide open at the neck, with a broad expanse of dingy-gray underwear showing underneath a medal on a chain. A wide-brimmed felt hat was pulled down to his large, long ears, and his welcoming smile revealed toothless gums below a hooked nose. He had a small, deeply lined face, with a day's growth of beard. I climbed out of the dinghy to the dock on my knees and when I stood up I was looking straight into his bright, dark eyes; we were the same height—five-two. He bowed low and removed his hat, revealing a bald head encircled by white hair.

"Hello, John, we see you come in today, and now you bring your lady to see old Frenchy and Simp*son*, yes?" he said, speaking with a pronounced French accent. He turned to me. "My name is really Leo Jacques, and sometimes they call me Jack, but people know me as Frenchy. My friend's name is Albert Simp*son*. You want to see Mr. Simp*son*, John?"

"We came to see you both, Frenchy," John said, as he straightened up from tying the dinghy to a rotting plank. He handed me the custard pie and chowder, and gave the spring salmon to Frenchy, then picked up the bag with the beer and whiskey that he had set out on the dock, tucking it under one arm and holding the bag of fruit with his other hand. He looked fondly at Frenchy. "You look very well—better than the last time I saw you. You were having trouble with your stomach."

"Boy, I wish I had a stomach like that old man Simp*son*," Frenchy exclaimed. "Oh, well, I am *cer*tainly O.K. and I do a lot of praying for Simp*son*. He's my friend." He started walking ahead of us, then stopped and said to me, "That old man, he looks like Chur*chill*. I'm eighty-four this month, but that man Simp*son* is eighty-eight." He held up the spring. "I'm sure glad to get fish. We haven't had any for long time."

The dock teetered drunkenly when we walked across it, and the plank connecting to the grassy shore quivered when I stepped on it. We passed rows of neatly piled firewood, uniformly cut alder, on the way up the path, and crossed the rotting porch. Frenchy opened a door, and we went into the house. It was so dark inside that it took my eyes a minute or two to get adjusted. We were in what must once have been a store, with a counter and shelves behind it just past the entrance. There was a calendar on the wall with a picture of a Japanese girl in a white kimono, and above it a green pennant that said, "Alaska, 1969." Farther along the wall was a magazine cover with a picture of de Gaulle, and above that a set of antlers was mounted on a board. An oil lamp hung from the ceiling, and several dark socks were draped along the sill of a dusty window.

"Come on, old man, see what company I brought you!" Frenchy shouted. I looked down the narrow, rectangular room in the direction he was shouting and saw a large, elderly gentleman wearing an old-fashioned brown fedora hat. He had a round moon of a face and looked absolutely like Winston Churchill. He was sitting up rigidly erect, smoking a cigarette in a holder held upright between two fingers that stuck straight out. He was dressed in a blue-and-white-checked flannel shirt and gray pants, and his feet, which were swollen, were encased in bedroom slippers. A crutch was hooked to one arm of his chair.

Frenchy drew up two chairs, then darted over to Simpson, leaned over him, and pulled up the open fly on his pants with the tender gesture one would reserve for a child. We sat down facing the big man, who was staring at me. Frenchy seated himself on a bed against the wall and mopped his brow. "I get so nair-vous and excited," he said apologetically. "Just with people I like. The others I don't give a damn about. I pay no attention." Simpson, meanwhile, had finished his cigarette and was fumbling to pick another out of the package he held, but he dropped the package on the floor. Frenchy jumped to pick it up, took out a cigarette, lit it, and handed it to him, saying to me, "This old man fell off a roof t'irty years ago and landed on his hands, so he broke them, and they aren't much use. He can't walk, neither. Since he had his stroke, he's partly—partly—" Frenchy paused, searching for the word.

"Paralyzed?" John suggested.

"Right you are, sir. That's the word for it." Simpson spoke for the first time, slowly, in clipped, precise English. His voice was unexpectedly deep and clear. "I have been living in this bay with Frenchy for fifty years, since we met and teamed up together." He turned his head toward me, paused to move his hand slowly to his mouth, and puffed on his cigarette. "I am of Scotch background, from Nova Scotia. We were a family of eight, and most of my family is around Boston now. I have a sister who writes often, and another sister who left a hundred thousand dollars to her church."

"The money we have, we decide to leave it to hospital," Frenchy announced.

"What hospital?" John asked.

"Oh well, whatever hospital we decide to die in," Simpson replied. "We split everything fifty-fifty. Frenchy's a smart man. He's a Frenchman, but he's honest."

"That Simp*son*, he's not a very good trapper, he's no good at all," Frenchy said. "He has trouble even catching a mouse!"

Simpson removed his cigarette between stiff fingers and gave a deep, easy laugh, revealing gums as toothless as Frenchy's. "Yes, I could learn a lot from Frenchy," he said.

"Have you been back to Nova Scotia often?" I asked.

"I never wished to go home again," he replied. "I worked first in Alberta, but I wanted to fish, so I came to B.C. and went trolling. Frenchy is from Trois Rivières, in Quebec. We used to troll together, but he's had traplines as well, trapping marten and, later, mink. He made five hundred dollars one month."

Simpson stopped talking, and Frenchy jumped up again. "Do you want a bottle of beer?" he shouted at Simpson, and Simpson nodded. Frenchy opened a bottle from the six-pack that John had put on the counter, and handed it to him. Simpson raised it clumsily between his two hands to his mouth, took a healthy swig, and managed to set it down on a table beside him. Frenchy offered us some, but John shook his head, so Frenchy poured us each a mug of coffee from a pot on a kerosene-drum stove under the window.

"I never went back to Quebec, neither," Frenchy volunteered. "My father was a doctor, and my mother died when I'm two year old, and he had a nurse take care of us children. My older brother had a big farm, but I left home very young and crossed the country. I never went to school. I got shot in the First World War." He opened his shirt. "See that hole?" He leaned over so we could get a good view of a small round scar. "The bullet went out back. And I got slit with a bayonet right across here." He ran his finger across his neck, then pulled back his shirt collar to show us that scar, too. "Only four or five come back, but, by jeez, this crazy Frenchman, he come back! I got a bunch of medals, but they sank in my boat."

"Show them the document that was presented to you with your King George medal," Simpson said.

Frenchy went to a drawer and took out a folded sheet of faded paper, which he handed to me. "Old King George and the Queen came to see me when I was pretty near dead in University Hospital, London," he said.

I opened the paper. "Read it out loud," John said.

Below the Royal Crest in red at the top, and the words "Buckingham Palace, 1918," I read, " 'The Queen and I wish you Godspeed, a safe return to the happiness and joy of home life, with an early restoration to health. A grateful Mother Country thanks you for faithful services.' " It was signed "George R.I."

Frenchy nodded proudly while I read. He pulled a card out of his shirt pocket and handed it to me. It was a holy card, with a picture of the head of Jesus wearing a crown of thorns, above a prayer in

French, printed in green ink. "I had this in my breast pocket when I got shot," Frenchy said. "I kept it from 1918, and I been all over the world with it on them freighters I used to go on—especially on the Great Lakes." He patted it and put it carefully back in his pocket.

"I had half my Army pay going to my sister, a very lovely woman, and when I came home I find she give two hundred dollars I send to help her priest to say prayers for me," Frenchy went on. "So I never send her more, and when I leave, I tell her, 'You won't hear from me, because you worry too much.' I never been back to see her—not since 1919—and I never wrote, neither. Not to my own sister."

It started to rain. Two or three raindrops hit my shoulder, so I slid my chair an inch or two back from a leak in the roof. Frenchy leaped up and put a tin can on the floor to catch the drips, and from then on, the patter of rain on the roof was accompanied by a musical *ping*. Frenchy sighed. "Someone brought this old fish camp here on a scow and left it all to rot. When we move in, she look like hell. I pull her up here and push some cedars under her when the tide is in. I patch the roof real good, but I guess I have to do it again. Last winter, John, we live real nice. Someone from Canadian Fish Company tow a fish camp with a store here, and we look after it, so we got to live on it until spring. Especially because it have an oil stove, it is much more comfortable than this old wreck. I have a lot of work carrying five-gallon drums of oil inside for the stove, but it's good all the same to wake up in a nice warm house, and we had very little snow." He laughed. "By golly, I sure never want to go back to Quebec. Too cold!"

A sudden crash. Simpson, lifting his cigarette to his mouth, had knocked over his beer bottle. Frenchy scrambled after it as it rolled away, and he muttered, "We have more trouble from morning to night." Without warning, he burst into song. "Ho! Ho! Ho!" Then he sang it in a faster rhythm—"Ho-ho-ho!"—smiling at Simpson. "We get some good laughs," he said.

"Yes, we do," said Simpson, with a deep chuckle. "We'll have the Seventh-Day Adventists in here soon and have a gospel night." We were all laughing now; Simpson's heavy body was shaking with mirth. "The United Church boat, the *Thomas Crosby IV*, stops in here every three weeks to see us," he said finally, taking out a blue bandanna and wiping his forehead.

"I know that boat," John said, still smiling. "It's a big steel one, eighty feet long, that goes anywhere in any kind of weather."

Frenchy went on singing little "Ho-hos" to himself as he skipped back to the bed where he had been sitting. He picked up a handsome patchwork quilt and a yellow-and-blue crocheted blanket there and brought them over for me to see. "We asked the *Thomas Crosby* to bring us sleeping bags," he said. "The one they gave us for Simpson was too small, so the United Church at Prince Rupert sent us these and I put them on this old man's bed instead."

"We had quilts like that in Nova Scotia," Simpson interjected. "Mother used to make them."

Frenchy ran to an alcove behind me, which was evidently his bunk, and came back with an armful of quilts, announcing, "I got two or t'ree more quilts that the Church brought us, and the five girls who crocheted blanket come to see us. One of them church girls give me a kimono, but I don't say not'ing." The quilts were handmade and beautiful.

Frenchy removed his old fedora, scratched his head, and put the hat on again, but Simpson never removed his for any reason. In fact, I never saw Simpson without a hat. (Later, when I asked John if he had ever seen Simpson without a hat, he said, "No, and nobody else I know has, either.")

Frenchy said, "We hire a guy to bring our groceries every four weeks—wintertime, too. He's a paperworker from the Falls, a Prussian with a nice boat that takes an hour and a half to go twenty-eight mile to the Falls. He do everything for us." He chuckled. "When I used to go to Ocean Falls, I used to go to the Happy House, right around the bend at what we call Pecker Point. The alders have *cer*tainly grown over that Happy House fast since it closed."

Simpson was frowning, shaking his head at Frenchy, who suddenly stopped talking.

Looking amused, John picked up the conversation. "That German you hire must have a powerful boat. It took us three hours to get here with my seventy-two-horsepower five-cylinder Gardner," he said.

Simpson leaned back in his chair, cackling. "Oh, I know that one. It's a good one."

"I used to have nice boat, troller, for seventeen years," Frenchy said. "The *Albatross*, built at Sointula, on Malcolm Island. In 1936, I get twenty-four cents a pound for springs and six cents a pound for coho in Namu—forty cents for a whole coho. A lot of people were living in this bay then, in small shacks in winter, mostly loggers and fishermen, but there was a woman here who sold baths. She had bathtubs in little huts, and you could take a bath for two bits or a dollar—whatever you thought it was worth. She would lend a towel to you for ten cents, or sell it to you for twenty-five; the only thing, she was such a bad cook. I know because I used to chop her wood to have a cup of tea and whatever she gave me to eat."

"What happened to the *Albatross*?" I asked.

"I sink it, I t'ink in '43, maybe '44," Frenchy said. "The manifold and exhaust and fumes, they put me to sleep and bingo! she went on the rocks. I get out in a rowboat, and someone from Bella Coola in a gillnetter came along and picked me up. No insurance, so I lost the whole works. I bought another boat the next year, and someone walked on it and stole eight hundred dollars." He shrugged. "Fishing from daybreak to night was too long, anyway. I got dizzy from the exhaust and from looking up at those lines."

He got out four glasses and poured us each a Scotch, handed them to us, and said, "There are some nice big ducks around and we don't even shot one. We like to see them around. We used to make home-brew and then t'row away the corn, and when ducks came around they'd get drunk and go round and round like this." He lay down on the floor on his side and flapped his arms, hopped up and sat down again on the bed, still talking. "They was fat and in good shape, and once in a while a hawk or owl came around and got one. We used to have lots of owls at night, but no more now."

"Every night, they used to howl 'Hoo! Hoo!' " Simpson said. "Another thing, we used to have a lot of Japanese here from the Falls. There was this old fellow who was eighty-four, a gillnetter, who used to hang around here in the winter. He wasn't a bad fellow, either. Better than that fellow that's over there now."

Frenchy leaned over and said to me in a low voice, "I don't mind, but him and Mr. Simp*son* don't get along well."

My eye had been attracted by a white ceramic angel with a gold halo on the shelf behind Simpson's impressive head, and when my glance shifted to binoculars hanging from a hook on the wall next to it, Frenchy said loudly, "From here, Mr. Simp*son* can see a bear way up at the head of the bay without glasses. He's got to do somet'ing, that guy. I shot two black bears last year. There used to be a lot of deer, too. They would come in evening at low water and drink, but I don't see none now. I fish for Simp*son*, but I wait until there is more than one fish in the water and then I don't make too much noise. Seals and otters come right up to your boat if they t'ink you have fish. They walk on the bottom, those otters—I see the tracks—and they say when seals come like that, fish are coming in. Seals walk one and a half mile to this lake above here, and then another mile to another lake. We saw twenty wolves in one bunch, too; they come down for fish. They howled right alongside here, right around our door."

Simpson signaled for more Scotch. Frenchy poured some into his glass. "Thank you, sir," Simpson said. He held the glass up between both hands and drank, put it down, and said, "We had the best cat in the country, and the wolves got him. Oh, he was a wonderful fellow."

"When anyone came, he just sat by Simp*son* and growled!" Frenchy said.

"Oh, the wolf caught him right at the door—I know he did," Simpson said mournfully.

"That black-and-white cat, Ta-puss, would sleep up on the pillow against Simp*son's* head and would snore," Frenchy said. "Simp*son* would wake up and say, 'Quit your snoring,' and push that cat to the wall. One night, he was sleeping right beside Mr. Simp*son*, snoring, and Mr. Simp*son* tried to push him off once too often and Ta-puss bit him in the hand." Frenchy began to laugh, rocking back and forth. He laughed so hard he had to wipe his eyes with his sleeve.

Simpson wiped his eyes, too, and said, "Ta-puss would push on the doorknob and then push the door open."

"Old Ta-puss, I wouldn't take fifty dollars for him," Frenchy said.

"I wouldn't take a *hundred* for him," Simpson said, shaking his large head solemnly from side to side. "Oh, I wish we had him now. He would never catch the mice in here. He would just sit and look at them walking back and forth in front of him."

The light coming in through the windows was fading, and the room had become even more shadowy. John rose to go. "Tomorrow, we're going to Nascall Bay. We'll stop in on our way back from there," he said to Simpson, who was looking downcast.

Frenchy walked back to the dock with us. It was now totally dark, and we rowed home in a driving rainstorm. Without my Stanfield's on, I was thoroughly chilled. It was so cold when we got back to the boat that John turned up the stove while we shivered, taking off our wet clothes. We ate supper surrounded by the smell of soggy clothes drying out on a clothesline that crisscrossed the cabin from wall to wall.

We were in bed long before ten o'clock, when the strike news came on. Nothing had been settled. John said gloomily, "It may be another week before the strike is over, but anyway it gave us a good chance to come here. I always stop and see those wonderful old fellows once a summer, even if it means losing a day's fishing. I sure admire Frenchy. It's not easy for a man his age to do all that cooking, and he washes their clothes, too, by hand. Frenchy tells me he has to lift Simpson sometimes; I don't see how he does it, Simpson is almost twice his size. No, no, Simpson is not an easy man to take care of. He liked you, by the way. If he hadn't, he would have sat there like a sphinx, not saying a word."

XIX

John moved the *MoreKelp* to a different position during the night, because of the wind, and just before we left the next morning, we looked out the window toward Frenchy and Simpson's place. With the tide out, we could see a neat gray speedboat tied up at the small dock that I hadn't noticed before because of the tall grass around it. John said, "It's Frenchy's and it's sixteen feet, with a little cabin and a gas engine. If their float house burned up, he could put Simpson on that boat."

Coming out of Eucott Bay, I saw a cormorant take off, skimming across the water and leaving a cascade of whitecaps and foam in his wake. The sun was gleaming on the mountains behind us as we turned left into Dean Channel. After a while, I climbed through the window where John was steering, and sat outside with my back against the pilothouse wall. He handed me my coffee. "Where are we going?" I asked.

"To the hot spring that's at Nascall Bay," he said. "I hardly ever get free time like this in summer, so we'd better take advantage of it. It's a sort of honeymoon for us."

"Hot spring! You never told me we'd be going to a hot spring," I exclaimed. I picked up my mug and scrambled back through the window. "You should have told me that before I came. I haven't got the right clothes for that. All I've got is bluejeans." The only hot springs I had ever seen were at Saratoga Springs, New York, and they had big hotels around them, with fancy dining rooms with white-coated waiters. It never occurred to me that there were other hot springs besides those connected with resorts.

John looked astonished, then laughed. "There's nothing like a good soak in a hot spring to remove the kinks from your system," he said.

"Well, what is your hot spring like, then?" I asked.

"You'll see."

We arrived at Nascall Bay at noon. A large seine boat was already there, so we anchored at the entrance of the bay—an even smaller one than Eucott—to wait our turn at the hot spring. We watched while four men with towels set off in a dinghy from the seiner, landed on the rocks, since there was no beach, and disappeared up the hill. John was looking through the glasses. "That boat is from Sointula, and they are probably Finns," he commented. "Sointula fishermen love hot springs."

In an hour, they came down again and went away, and before the seiner had disappeared we were rowing in our dinghy for those same rocks. John tied the painter around a large boulder, and we climbed the rocks up the hill. We arrived shortly at a small, rough-board cabin with a peaked planked roof, which had been set up across flat rocks around the hot spring in a woodland setting overlooking the sea. So much for my fantasies about fancy hotel resorts and fancy dining rooms with white-coated waiters!

The cabin was just big enough to contain a bathtub and a wooden bench. There was a white-painted doorframe, and someone had hung a long piece of weather-stained canvas there for a door and white sheeting as curtains over the empty window frames. The words TURKEY TOWN II were painted in foot-high red letters in a slant over the door, and below the front window, SHOWTIME LODGE. Someone else had scrawled RELAXO! in giant letters vertically down the front wall to the right of the door. Inside was an ordinary old-fashioned porcelain bath-tub, sunk in the floor, into which steaming springwater ran in and out in a steady stream through a pipe. John swished the running water around to rinse out the tub, then put a chained plug into the drain, and the tub gradually filled up while he took off his clothes and laid them on the bench. "It's probably about a hundred and forty degrees Fahrenheit," he said. "In the wintertime, hand loggers used to throw armfuls of snow in to cool it down. We'll have to wait a few minutes." I leaned over and felt the water. It was good and hot, and had a sulfury smell.

John stuffed a rag hanging over the pipe into it to stop the flow of water, and waited a bit longer. He felt the water, climbed into the tub, and sat down. After he had finished washing and soaking, he opened the pipe again and held his wrists directly under fresh water coming in. "God, this feels good," he said. "Wonderful for my arthritis. I wish this tub were big enough for both of us."

I sat on the bench and waited, and when my turn came, the water seemed pleasantly warm instead of being hot enough to cook in. No other boats arrived, so John fastened the canvas door above the jamb, and I could look out on the glistening blue water of the bay while I floated in my medicinal warm bath, the air fresh and cool around me.

"What are you thinking about?" John asked.

"I'm thinking: What am I doing on this rocky escarpment, sitting in a bathtub sunk in the ground—you might almost call it a mudhole—staring goggle-eyed at a gorgeous view, in the middle of a misty-green wilderness that has the quality of a Chinese landscape painting?" I said.

"You are here because you're with me," John said, smiling and looking pleased with himself. "How do you like our hot spring?"

"Paradise," I said.

I sat in the tub for quite a while, with steam rising around my head, the sour smell of sulfur in my nostrils, and the water almost up to my shoulders. A little groggy, I wiped beads of perspiration from my forehead. "I feel a bit dizzy," I said.

"Time to get out," John said. "You've had enough."

The interior walls of the cabin were covered with carved and written names and dates of previous bathers, going back thirty years, and John and I added our own initials and the date before we left. As we rowed in the dinghy back to our boat, John said, "There are hot springs up and down this coast, on the west coast of Vancouver Island and in the Queen Charlotte Islands as well. The Indians have always known where they are, and fishermen, loggers, and the locals have used them regularly for years and years. Someone might come along and install a tub or build a concrete pool around one, and another fellow, usually a logger or a fisherman, might come by and put a shelter over it, and then another guy would pass by and improve the roof or repair the pipe or install a better spout."

When we arrived back in Eucott, a handsome new white troller, the *Salmon Stalker,* with a freshly painted light-green pilothouse, was already anchored at the entrance to the bay. The owner, Al Perkis, a big man with a large black mustache, had his young son with him, and they greeted us from a gray rubber collapsible skiff with a six-horsepower motor attached. They had fishing poles and were about to take off for a little late-afternoon trip around the area. Perkis was an old friend of John's, and lived in Duncan, on Vancouver Island, near where John had grown up. "I've been doing a little trout fishing during the strike," Perkis said. "Right now, I've just been to Bella Coola, where I've been looking for land. Duncan, B.C., is getting too crowded for me."

"I know what you mean," John replied. "I've been thinking of doing the same thing myself. How's Jonesy?"

"Right now, Jonesy's sitting in Masset in the Queen Charlotte Islands," Perkis answered. "He'll be down soon if this strike keeps up much longer."

They left, and, as we watched them go, John said, "Perkis fishes a lot with Gordon Jones, each on his own boat. Their boats are fairly small, and both are tiptop fishermen, terrific producers. Jonesy is one of those uproarious bachelors who don't like towns, even though he lives in Duncan, too. They are both absolutely nuts over steelhead fishing. It's hard fishing—fighting those big trout that spend a lot of their life in salt water. And the steelhead trout is unique, one of the most sought-after fish in sportfishing. Both men do all the right things with their boats—copper painting and so on—but the main thing is that they do it *all* by the end of September, because in October, as soon as the rivers are high, they start fishing steelhead, just for the sport of it, for fun. They use barbless hooks and play the steelheads and release them, and compete to see who can catch the most steelhead. Jonesy once got a hundred and twenty. It goes on all winter, so you can see how they wouldn't get much work done on their boats then. Whoever is ahead has to buy the other a case of whiskey, and they usually don't get more than a fish or two ahead of one another. There was a last day of steelhead fishing recently that I heard about where they were even. They started off in a pouring rain and had the case of whiskey in someone's truck that followed them on a railroad trestle across the river, where each was fishing on one side. The fish moved across, and they both hesitated, and one made a mad dash and the fish kept moving and so did they. I don't know who won, but they really are the nuttiest fishermen in the world."

Perkis and his son arrived back in the early evening while we were eating supper, without having caught any fish. Meanwhile, the weather had turned unexpectedly warm and we were being attacked by horseflies as big as bumblebees. After we turned out the lights, they buzzed around all night. The *Salmon Stalker* was gone before dawn, and as early as we could get there on the tide we rowed over to say goodbye to Frenchy and Simpson. Frenchy was on the dock to greet us and hurried us in to see Simpson. "I cooked that spring last night, and tonight we have a hash wit' onions and potatoes," he said. "We sure are glad to see you. Mr. Simp*son* had me up all night watching out for you."

"Right you are!" Simpson said. "I told Frenchy, 'I don't see no light. You better go and see if they are all right, or I won't sleep very good.' One slip, you know, and that's the end of you."

"So I went out with my boat and row over to make sure you were all right," Frenchy said. At John's look of surprise, he added, "I'm very quiet, so you don't hear me."

We left shortly, and Frenchy came along beside us in his motorboat, carrying a big can and several pails to get fresh spring water from a waterfall in the next bay. We went with him to the freshwater spring; John helped him with the heavy containers, and got some sparkling-clear drinking water for us, too. We were in Cascade Inlet, a long, narrow body of water which had eight waterfalls marked on the chart. In this bay of many waterfalls, we ran past four beautiful ones tumbling over steep cliffs in clouds of spray. "Waterfall chasing is a favorite occupation of mine," John said, lingering at each one. When we went out, we saw Frenchy moving at terrific speed, rounding the bend in the opposite direction, going home. We went back to visit Frenchy and Simpson in Eucott Bay whenever we could. The last year John and I fished, when we rowed over to see them, Frenchy came out to meet us with the news that Simpson had died that winter. We went inside and had a cup of coffee with him. He was a heartbroken man. "Rheumatism just ate Mr. Simp*son* up," he said, wiping his eyes. "When he was dying, he ask me to pray for him, which I did, for half an hour in the middle of night. And then Simp*son* said, 'Now I feel better.' I have to pick him up, and light cigarettes for him at 2 a.m., and he start fires in his bed. I have to send him to hospital, finally, in Ocean Falls, and after t'ree days he died. I was so exhausted I didn't go see him, even though he ask for me." Frenchy put his hand to his eyes again. "I would have cried, so I just stayed here."

After that first winter alone in Eucott Bay, Frenchy moved to Ocean Falls, boarding with friends until he became too old and infirm to look after himself, as he had so lovingly done for Mr. Simp*son*. He spent his last days in the hospital at Bella Bella. He was over ninety when he died.

XX

After talking to Al Perkis, John thought perhaps we, too, should explore the possibility of moving to Bella Coola, so we went there. Bella Coola, located at the end of another long, winding waterway nearby, is a settlement known for the beauty of its mountainous surroundings, its intemperate climate, and its isolation—it can be reached overland by a single, homemade, almost perpendicular road, suitable only for mountain goats. We sat at the Bella Coola government dock for several days in a drenching downpour; the continual rasp of our boat rocking against its neighbor, the rumble of the Gardner being run to prevent the chilly damp from overwhelming us, and the endless patter of raindrops on our roof got on our nerves. We were packing up to travel—stowing loose items in the sink, ramming the knife in the cupboard door, and so on—when the fisherman on the next-door boat told us that a strike vote had been called for the following day. We departed at once for Namu.

At slack tide, in Fitz Hugh Sound, John slowed down to put one line out for a fish. Fifteen minutes later, he caught a small coho, dropped it in the fish box, and pulled up the pole; then we continued our journey. He looked doubtfully at the fish, measured it against the crisscross lines at the edge of the box, murmured, "It's a little undersized," and hastily cut off its head. He hurriedly dropped it on a shelf under the cockpit. Black clouds were forming in the sky, and, as the weather worsened, he said uneasily, "We'll just leave it there until we have a chance to cook it."

In a gray, wild sea, we passed a large combination troller and gillnetter with halibut flags hanging off the stern, which had a picket sign on the boom. The vessel turned around and cut us off. We stopped and maneuvered as close to it as we safely could. Three large men had come out in the stern, and one of them leaned over the side of the big vessel and shouted, "What are you doing here?"

"We have a permit to travel," John shouted back. "We did picket duty at Bella Bella, and we're on our way back to Namu from Bella Coola."

"That's O.K.," the man said, and went on, "It looks like a long strike. I heard that the shore workers accepted a one-year contract, but we fishermen want a two-year one and won't settle."

John relayed the news to them that a strike vote had been called. "I'm on my way to Namu to vote," he shouted.

"We're from Sointula," the man said, as the boats moved away from one another. "Striking is a bugger, but we're only trying to get enough to live, aren't we?" I could hear him telling John the price his wife had to pay for bacon and pork chops, but the wind was whistling, so I couldn't catch the amounts. I heard John shout back that the price of radar equipment had doubled in the past three years, but we had moved too far apart in rough water to talk further. We proceeded on our way, and I saw the big vessel turn around and follow us.

"Sointula is seventy-five percent Finnish," John said, when we were settled down again in the pilothouse. "Sointula men are the most terrific workers—huge men who eat and drink like Henry VIII. I couldn't do what they do; I'd have a heart attack. They are either far to the right and money-conscious—their favorite saying is 'A man's best friend is the dollar'—or far to the left. Very fine people who are remnants of a utopian cooperative community started there around 1910. It failed financially, but some of the original settlers remained, and they are among the most skillful loggers, boatbuilders, and fishermen in B.C. One of the Sointula boats is called *The Millionaire*. Left or right, they are all *nuts* about the island, and a lot of people have moved back there from Vancouver and bought farms. It sure helps there to be Finnish. Their parties are absolutely famous. They go on anywhere from two days to a week of continual dancing, drinking, and eating. Sointula is also renowned for its good-looking blond women."

We passed a gillnet fisherman from the Native Brotherhood, winding in his net. Increasingly, B.C. native fishermen have asserted aboriginal rights to historic fishing grounds, opposing government regulation and other efforts to control their freedom to fish in their own way. I asked

John what mistakes he thought had been made in dealing with native fishermen, and he said crossly, "Mistakes, mistakes. Life's nothing but mistakes. We conduct a struggling course through the mistakes. I want to write one book before I die, called *Mistakes.*"

A white fiberglass speed cruiser on our port side was heading directly for us, and I reported this to John. "Good God! That must be a government Fisheries officer!" John exclaimed. "He must have seen me, with his glasses, taking that undersized fish." John stopped as the white cruiser cut across our bow, and it came alongside, idling its motor to match ours. On the side of the vessel was a sign, ENVIRONMENT CANADA, and a young man who was running the boat from the upper bridge leaned down and shouted, "Have you seen any killer whales?"

"None so far around here," John shouted back.

"If you do see any, will you let us know?" the youth called out. "I'm photographing them, especially for their fins, for the government." He departed at high speed, heading for the seiner behind us.

"Let's go in and cook that salmon and eat it before we get into any more trouble," John said. So we did.

When we arrived in Namu, John tied up again alongside Jimmy Chambers. There was a feeling of heaviness in the atmosphere around the floats now. The fishing vessels, their holds empty, were riding high in the water—so high that you could see where the copper paint started on their hulls above the waterline. You could hear their sides creaking against one another and the floats—a constant grinding noise as you stepped carefully from buoyant float section to float section. Everything in Namu was closed up: the ice machine was still, the loading equipment motionless, the piers and working docks empty; even the ravens seemed listless—the dogs who were their favorite dive-bombing targets had disappeared. The floats were full, packed with the idle boats of the fishing fleet, four and five deep, but there was an unnatural quiet among the fishermen. They sat around in small clumps, or leaned over the net racks on the floats, arguing in low voices. "I just feel fed up," Jimmy said. "I don't feel like talking to anybody."

A thick fog rolled in, increasing the gloom. Across from us and down farther on our float were also boats of nonunion fishermen. "They're grounded, too," John explained. "The whole industry's tied up in this fishing strike."

John went off to a union meeting and I walked up the hill on the path leading toward the lake, to visit Margaret Harrod. Even her bright spirits had been dampened by the events. When I returned to the float, John had finished voting. Union members were spread out along the entire B.C. coast, so we probably wouldn't know the result for another day.

As soon as I arrived, John took off for Boom Bay, which was beginning to feel like home. The fog had partially lifted, but the air had a sharp chill, and the bay, as we entered, was swathed in deep mist.

An old boat, with a small cabin, was lying among the wreckage of a dock and piled-up timbers along the shore, and while we were eating supper on deck, I asked John if he knew where it had come from. "It's been here a long time," John said. "Someone was telling me today that while he was fishing recently he saw a boat up on the beach that had been smashed almost to pieces. He and a friend went over in a rowboat and found a fisherman, a Russian who was about seventy-eight, dead in bed from a heart attack. The boat had gone up on the rocks and was a total loss." He shrugged. "Too bad about the boat, but that fisherman probably went the way he wanted to go."

"You told me once that you had five boats. When did you get your first one?"

"When I worked for B.C. Electric in Vancouver, I used to take my sandwiches down to the harbor in my lunch hour and watch the boats," he replied. "When my mother died in 1933, I inherited a small amount of money, so the next year I quit my job and bought a thirty-two-foot gas-engine boat from a man named Alex Znotin for twelve hundred dollars and began fishing. The boat had had different names—*The Baby*, for one, and I changed it to *The Snark*. It was two feet narrower than this boat, with an eight-foot beam, and it had a little doghouse on the deck which I steered from, sitting right above the engine and exhaust pipe. I did my cooking and sleeping below. It had a fishhold so tiny I had to dive in to get down there, and a bunk I had to double over and slide into. It was so small down below that not much cooking was done there. I kept *The Snark* until 1937, and then I sold it and bought my second boat from Big Alex. Znotin was a huge man, a White Russian who had been a naval officer on the Baltic Sea. He left Russia during the Revolution, came out to Canada, and worked as a mate and winch man on freighters here and lived on Gambier Island, in Howe Sound. He was the son of a ship chandler, and he used to pick up ship hulls and repair them. He learned enough to build a boat of his own, which he had for two years, and then I bought it. It was thirty-six feet, with a nine-foot beam. He called it the *Maureen*, after the daughter of the man with whom he boarded, and I changed it to the *Muriel D.*, after my mother. It's now called the *Alexandrian*, and it's the strongest-built boat for its size in B.C., but it's a horrible sea boat. It must have been hard work to build. The boatbuilders used to go into the woods and find yellow cedar with natural crooks that they called knees, where the roots joined the trunk or where the tree trunk conformed to the curve of a hillside, and attach these inside the hull in a lot of boats, instead of the metal clamps used to join solid timbers at right angles now. The *Alexandrian* was recently up for sale in the union newspaper and I think that by now it must have been sold again. I never worried when I had it, because everyone said it would be impossible to sink, but I wanted more capacity, so I got the *Muriel D. II.*"

"What happened to *The Snark*? Is it still around?" I asked.

"I've lost track of *The Snark*," he said. "I saw it last in 1953; it had a standard pilothouse by then, but I recognized the hull. I went out on a party with the young fellow who had it at Bamfield, on the west coast of Vancouver Island. We stood on the float and sang 'Goodnight, Irene,' and he fell in the water on the last verse and I had to pick him out. It was a foggy night like this, and damn cold water. His wife was not very pleased."

"Did you buy the *Muriel D. II* from Alex Znotin?" I asked.

He smiled. "No. I had that one built by Kuramoto, the Japanese owner of the KM Shipyards in Vancouver. He was interned right after Pearl Harbor. That boat cost twenty-four hundred dollars, including the fuel tanks. He went overboard to give you your dollar's worth; I found him super-duper reliable and unbelievably honest. He lived to be ninety-two, and I still keep up with his sons. One is about sixty and gillnetting, and the other became a hermit in the woods at Lonesome Lake, in the Bella Coola Valley."

"That makes three boats. What about the two others?"

"I had my third boat, the *Muriel D. II*, from 1940 to 1945," John said. "It was during the Second World War, and by then I was married, with a family, and up to my neck in a commercial peat-moss enterprise. It was made a war industry, which practically put us out of business. One of my two partners still has the company and has done very well with peat moss, which is used all the time now in horticulture, but I lost a lot of the dollars I had inherited from my mother in that business, and I hated it besides. I couldn't pull out, because my partners had their money in it as well as mine, and one was in the Air Force and the other in the Army, so there was nobody else to run it. The upkeep on the boat was killing me, and I had to charter it to others: to my friend Charlie Walcott, for fishing; to the Finning Tractor Company, to do service work; and to a guy who went dogfishing. Finally, I sold the *Muriel D. II*, but I didn't stay without a boat very long—just a couple of months." He paused, looking thoughtful. "That was *the only* time I've been without a boat since I got my first one. Then I bought boat number four, the *Kelp*; thirty-four feet long, with a nine-foot beam. Its name was always *Kelp*. The fellow who built it, in the late thirties, was scared of it; it was narrow and long, and he thought it would turn over, so he sold it to Charlie Walcott, who put in a lot of ballast. Charlie was a big, booming man, a hard drinker and a hard driver who would go anywhere with his boat, even though he had had polio and was a paraplegic. He had ropes rigged up overhead all over the boat. He could move around like a monkey hand over hand, using his powerful arms and shoulders, as fast as I can walk, maybe faster, and drop off the boom into the cockpit to fish. I wish you could have seen the way he swung himself about on that boat. He always took a youth or two as crew, and in the five or six years he had the *Kelp* he made a lot of money with it."

John began gathering up the supper dishes. "I bought the *Kelp* in 1946, and sold it in the fall of 1951 to a friend for four thousand dollars," he said. "*This* boat, my fifth, was built right at home, the next year, by a very good boatbuilder, Fred Crosby. I saw a design by a Seattle man named William Garden and bought the plans from him for three hundred dollars. He always put a man standing in the cockpit smoking a pipe in the drawings. The design was for a much bigger boat, and we cut it down, because I didn't think I could support such a large vessel. I wish I hadn't gotten scared, but I had had some poor years and I was afraid to commit myself too far. It cost ten thousand two hundred dollars, with the Chrysler engine. I was in a float house having dinner with another fisherman while it was being built and I said, 'I don't like the name *Kelp II*,' but I don't know what to call my new boat. His wife said, 'It's simple. You've got *Kelp*. Call the next one *MoreKelp*.' "

We spent the evening getting the gear ready for fishing again. John had been urging me to learn how to prepare the leaders he attached to the mainlines; he said it would be the greatest help to him if I did. I have always enjoyed swimming, playing tennis, and horseback riding, but I am not well coordinated or mechanically minded. It is one thing to be cheerfully clumsy for one's own pleasure but quite another to make a mess of someone else's serious work, so I was very timid about taking part in any important step of actual fishing. My reasoning was: John's been doing everything required to fish with great competence without me for over three decades. Why court disaster? On the other hand, there was no one who knew better than I how tired he was after he spent from ten to fourteen hours in the cockpit and had to rig up a new set of leaders for the next day's fishing before he went to bed. I had ignored whopping hints that I could ice fish; going down into the hold with all those dead fish had given me a mild claustrophobia. In the end, I said, "Yes, I'll try making up leaders, if you don't get mad at me if I have trouble tying everything together, or something awful happens. You should know that I have never been able to tie a proper knot."

John was delighted. "Nonsense!" he said. "Nothing to it! You'll master it in no time." He gathered all the parts that make up a leader on his bunk, beside the big roll of clear monofilament line that always hung on the post beside it. He laid out the boxes of small quick-fasteners, greeny-yellow hoochies, little swivels, and black hooks, and handed me a set of pliers with red handles. He showed me how to clip off a measure of line, tie the fastener to one end, thread on a hoochy, and attach a swivel and hook below it, tying each knot tight with the pliers, which I could do if I gave a little jump and yanked simultaneously on the ends, bringing my full weight down behind the jerk of my arms to make it fast. The monofilament had to be wound around in and out, over and under, looped and twisted, in a very special way. The minute John walked away and said, "Now do it by yourself," I panicked. I twisted and looped, and the knots didn't hold. We had to keep going back over the steps dozens of times. "I'm thinking positive, and so must you," John would say with an affectionate smile, and start all over again. By the end of the evening, I had a neat pile of coiled leaders ready for him, each held in place with a clothespin, but I was worn out.

XXI

At dawn the next day, John started the Gardner, and we began moving. I got dressed in a hurry, but by the time I had scrambled up into the pilothouse we had stopped again. I looked out to see where we were. We were in Namu, at the chute of the ice plant, with one boat ahead of us directly under it.

"I'm usually first," John said, looking annoyed. "Whenever fishing starts, there'll be an awful crush to get ice, and I don't want to wait around." We didn't, either. Word came that the strike was over. As soon as we heard that, we collected our four tons of ice and were gone.

When John started fishing and came in for our main breakfast, he was unusually quiet as he sat down. While he spooned his soft-boiled egg out of its glass cup, I asked, "How's the fishing?"

He silently continued eating his egg. "I caught a sixteen-pound spring," he said finally.

"Wonderful!" I exclaimed happily. "What a great way to begin the day!"

"I caught it and then I lost it," he said. "I was bringing it in and leaned down to gaff it while I held the line up, and it got away. Hook and all."

"How awful!" I said. "Does that happen often?"

"No," he said.

Still it didn't dawn on me. Not until he came in half an hour later holding a bunch of leaders in his hand. "I just lost a fourteen-pounder," he said, looking bleak. "That's two mild-cure in a row."

I burst into tears. Then I reminded him about me and knots. "It's some kind of a hang-up from my childhood," I said. "I left the Girl Scouts because I couldn't tie knots. Think of something else I can do." It was his turn to comfort me.

That evening, for supper, he cooked fresh abalone that he had bought in Namu as a surprise for me from the woman skipper of a sailboat. He served it with a bottle of bubbly white wine he had been saving, and then proposed I take further instruction on leaders. I protested, and he was adamant. That evening, I was given a second lesson, this one exclusively on winding and twisting the vital knot that holds the hook to the leader. Whether it was his persistence or abject fright that cured me, I'll never know. Whichever it was, he never lost another fish because a knot of mine slipped; or, if he did, he never told me.

We went back into our usual pattern of fishing, and I lost track of days of the week or the month we were in. The only true measure of time was in the fishhold. I had to reach farther and farther down to get any meat or butter buried in ice—we had long since run out of fresh milk or cream—and, finally, to climb down as the melting-ice level dropped below arm's reach. I used an increasing length of rope to pull up and let down the white box containing our dwindling grocery supply. Whenever John was running out of pen boards to support the iced fish, which rose layer upon layer each fishing day, and the surrounding ice became watery, it was time to go in, to Namu or Port Hardy, to deliver our fish—anywhere from eight to twelve days from the time we started out.

We fished our way down Fitz Hugh Sound along the east coast of Calvert Island, and one fine day, as the sun was setting in a vermilion sky, John announced that we were going to a place called Finn Bay, to visit a fisherman named Ken Moore. "He lives alone in a float house, with a canary and a lemon tree," John said. "Ken's not a very bold fisherman. If he fishes in the next bay, that's his deep-sea operation."

We soon turned a corner where tall evergreen trees stood as sentinels above an almost vertical brown embankment, at the mouth of a narrow bay. A short way in, on our starboard side, was a collection of neat white buildings with pitched roofs, all fastened, on floats, in a line to the shore. In the center was a cottage with a sun-room full of windows at one side—the kind of small, neat house you might expect to find on land in one of the older suburban communities. "It always reminds me of a train," John said. "Every time a logger left, Ken would attach that logger's house to the others he already had."

On the side of the building closest to us as we came in was a sign: PRIVATE FLOAT. NO MOORING. KEEP OFF. John tied his boat to Ken's combination troller and gillnet boat, which was alongside a small dock, and went into the house, coming out shortly to beckon me to join him. I stopped on the way to smell the fragrant yellow roses that were climbing the outside wall of the cottage. "Take a look at the garden up there," John said, nodding toward the hill alongside. It was filled with bright flowers and neat rows of vegetables all the way to the top, where a homemade greenhouse was standing.

Ken Moore, a brown-haired man of about fifty and of medium height, with large eyes in a smooth, round face, was standing at the door. He solemnly nodded at my admiration for his gardening, and led the way into the house through the glassed-in sun-room, whose principal occupant was a large lemon tree filled with waxy white blossoms that threw off a delicious scent. He immediately started the teakettle humming. We sat down in a little nook at the kitchen window, overlooking his garden and a tiny lagoon that lay between the float house and the shore. The walls of the spotless kitchen and the room beyond were painted a glossy cream color, with green trim on the windows and cupboards. It was unusually cozy and cheerful.

"You look a *lot* thinner than you were when I last saw you," John remarked.

Ken appeared very pleased. "I've lost forty pounds in six weeks," he said, and sighed. "I've had to give up everything else to keep on eating two chocolate bars a day and all the peanut butter I want."

I complimented him on his house, and he got up and brought an armful of hooked rugs with large floral patterns for me to see. "Someone brought me a catalogue, and that started me off," he said. "The four rugs I've got here each cost over a hundred dollars to make, not counting the labor, and I've given as many away."

He noticed that I was beginning to shiver, although the oil stove was on, so he went over and shut the door between the kitchen and the sun-room. "That lemon tree freezes me out. I have to leave this door open to give it enough warmth from the house. I started this tree from a cutting from my previous lemon tree, which was so big I had to give it away."

"This is the nicest float house I've ever seen," I said.

He scowled. "I hate floats," he said. "I wish I could plunk this house on a hill." He was interrupted by a peculiar, rapid tattoo sound. In a bird cage hanging in the next window, an orange canary with a deep-brown feathered head was banging its beak against the bars. Ken went over and tapped gently on the cage. As soon as we looked up at the bird, it began to sing. I had never heard anything like it. The canary filled that room with its song, as it trilled and trilled in liquid tones. When it stopped, there was a vacuum.

"He was a present from some young men whose place I looked after who knew I like birds," Ken said. "For the first year, he wouldn't sing or bathe, but now he sometimes takes two baths a day. For two years, he was so scared he wouldn't go near me, but now I can put my finger in his cage and stroke him. He's lost when he comes out, so I don't usually let him out." He gazed at him with obvious affection. "I wish I didn't have him," he said. "I don't trust anyone to leave him with, and he's a bit of trouble, so I don't go very far away. I fall to pieces in Vancouver anyway, so I never go there. I recently flew a man up to fix my boat engine so I wouldn't have to go to town." He stopped, then said, "Just a minute, I'll be right back."

He dashed outside, returning with several lettuce leaves, which he stuffed through the bars of the bird cage. "I forgot to give him his dessert today," he said. The canary raised its head high and treated us to another golden song.

Ken shifted in his seat to get a better look at the bird. "When I was a boy, my only interest was birds," he said. "I had a pet thrush I used to feed, and a heron called Angus, here. We all used to catch anchovies to feed Angus; he ate eighty a day. I remember an eagle once appeared in one of the gillnets, and everyone rushed out to see it. It was injured and we saw to its recovery in this bay, but you know we get an awful lot of sportfishermen here. One of them, an American potato king who showed up, shot that eagle and used it for bait."

We invited Ken to have supper with us on the boat. He said that he had already eaten but that he would sit with us while we ate. On the way through his sun-room, he showed me a baby tomato plant, a pot of rosemary, and a large gloxinia, all thriving. Outside, he ran up the hill, and I could see him picking in his garden. Returning, he presented me with a bunch of parsley, a half-dozen carnations, and a soft, dewy, pink rose, a climber he called Little Bess, which resembled my favorite rose, New Dawn. He fished a long root that looked like a white radish out of his pocket and gave it to me. "I'm growing Belgian endive," he said. "Put it in sand and let it head up again. The forced head will be the endive that sells in the markets for over two dollars a pound."

When we passed one of the small white outbuildings lined up with the house, Ken said, "I've been here twenty-eight years, and that used to be a store I had for a while. I also sold oil and gas, but I quit to go fishing." Later, when Ken sat with us drinking tea while we ate spring salmon, he said, "I can't catch a mild-cure, no matter what. The others go out and catch them, but I get all wound up when I'm trolling and can't stop and I just can't catch a spring over eight pounds."

"The first three years, I was the same way before I'd catch a spring," John said sympathetically. "What about your engine? Does your boat thump? I had a generator once that used to rumble and I got some sponge rubber to deaden the noise. Do you have a singing wheel?"

"Maybe," Ken replied. "I've had two boats, and the same wheel on both. Do you think noise goes down the lines instead of through the wheel?"

Supper over, the two men climbed into Ken's boat to examine the noise problem, turning on his engine and disappearing below. When they returned, Ken was carrying one of the black boxes that contained killer-whale recordings to discourage sea lions from coming around fishing boats—the one designed by Fred Welland at Namu for underwater use. Ken turned it on, and a succession of thin shrieks and roars was emitted from the black box, followed by a spooky wail.

Ken turned the box off, and looked around the pilothouse. "I notice you don't have radar," he said.

"One of the reasons is, I don't know where to put it," John said.

"I think radar increases the electricity on a boat just about three times," Ken said. "I think it affects the rigging in this wet climate even though it doesn't touch it."

After Ken left, we followed with our eyes his light moving around in his house, until it settled in one place and gave a warm glow in the gathering darkness. We stood on deck and watched little ducks gliding by and gulls flying so low we imagined we could hear their wings moving. A coho jumped in a glistening arc through the air, followed by stillness. We lay down on the deck, looking up at the sky, and enjoyed the starry stillness until it was time to go inside for the weather report on the radio.

Ken came down to see us in the morning, carrying six cans of his home-smoked salmon and a side of smoked fresh coho. John had presented him with a bottle of Scotch when we arrived, and I gave him some paperback books I thought he would enjoy. He looked them over carefully, and then handed three of them to me. "You can have these back," he said. "I never read books written by women."

XXII

The days slipped by. We were preoccupied with John's fishing and, to a lesser extent, with my writing. If I lingered too long talking to John, or just watching, which I liked to do, he would tell me to go in and get to work. At that moment, I was writing about an architect, and I don't think I've ever known anyone less interested in architecture than John was. "I don't care about buildings, and I spend as little time as possible in them" was the way he summed up the subject. However, what he built in our house—a mantel over the handsome stone fireplace, shelves in the living room, towel racks in the bathroom, and a spice rack for me in the kitchen—were well constructed, with graceful lines; and always of pencil cedar, a fragrant and rich-colored wood that he got from trees he nurtured on his land.

Whenever I could, I would read to him what I had written that day. He didn't seem to mind listening if he thought I needed him, and he was a good listener, with an excellent ear for the sounds of words. He was an omnivorous reader, who loved good writing, and most suggestions he offered made sense.

Around that time, I decided that I should see what it was like to actually fish, so one sunny afternoon, with a hint of autumn chill in the air, I dropped down beside John behind the fish box to try it. First, John brought a box for me to stand on, to make up for the fourteen-inch difference in our height, so I could operate the gear. Then he tried to show me how to use a gurdy and let a line in and out. His fishing technique looked smooth and simple when I observed it from the other side of the fish box, but, of course, it was a different story to work a line myself. To attach the leaders with their lures and hooks into the mainline, I had to watch for the periodic appearance of the fathom markers to which I must clip them, while shifting the gurdy-control handle to pull the line in and out. Coordinating these moves was terrible, and I had to keep bringing the line back to a marker I had missed. There was a breathless moment when we thought I would drop a pig while I was fastening it to the line. I caught it just in time.

"I know enough *now*!" I said to John. "Can't I stop?"

"You have to catch a fish first," John said.

It thrilled me to look down into the water and see the silvery body of a salmon gliding along beside us, and realize he was attached to the hook on my line; but my attitude was all wrong, because I simultaneously felt very sad. With John leaning over me, giving instructions, I managed to pull it in, but I couldn't kill it; he had to do that. When I brought in the salmon, a coho, John said, "Take the hook out of its mouth," and handed me the pliers. I removed the hook, after a few clumsy false tries, and dropped the fish into one of the checkers in the fish box. I had to finish bringing in that line and the pig, catching three salmon in all. Then I cleaned a fish for the first time in my life. I had seen John do it hundreds of times, but the smooth circular motion with which he removed the gills eluded me; I literally sawed the gills out with the knife, and then I slit open the belly, removed the entrails, scraped out the cavity with the knife's spoon end, and washed the fish down. I placed it in one of the checkers in the fish box, cleaned the two other fish, put them on top of one another, and covered them with burlap. I threw the offal to the seagulls, watching them catch it in midair, listening to their shrill cries. There was an odd unity in all this; I was being John—part of the natural cycle. The wind and spray rose around me. I looked down from the stern and saw the foam being churned up by the propeller; I could feel the power of our vessel surging through the water; but I had had enough. "That settles it," I said when I was safely on the other side of the fish box. "It's you I love, and not the fishing." I never wanted to fish again, and I never did.

That same afternoon, while we were listening to the concert hour with the gear up, a large fish packer went by us. John picked up the glasses and watched it pass. "I did some collecting in the *Muriel D.* in the nineteen-forties," he said. "I was so far in debt to the peat-moss business and the boat, I wanted to be certain of getting some income, and it was sure money. The manager of the cannery I was working for was with me, and we went all up and down the Skeena River and to the Queen Charlotte Islands chasing seine boats. That was before I bought my place at Garden Bay, in Pender Harbour."

"When did you move there?" I asked.

"In 1947. But I bought the land two years before that," John said. "My neighbor, Fred Claydon, who is in his eighties and lives in Victoria now, found it for me. We met when we were both trolling."

The music program ended, and John started putting out gear, continuing to talk. "Fred and I were coming south on our boats together, and he said to me, 'If you want some land, how about a chunk of mine? A quarter of a section is too bloody much land for me. I paid two thousand dollars for a hundred and sixty-two acres, and I'll sell you half, for a thousand dollars.' " John smiled. "I only had four hundred dollars in cash, but I had six hundred dollars' worth of lead, and he was building a boat. He took the six hundred in lead, and I paid him the rest in cash. We struck the line of the land right up the mountain, where it ends in a gully. He took the level chunk because he has a wooden leg; it was easier for him to get around on the flat. Later, I asked Fred if he minded if I subdivided, and he said, 'No, go right ahead.' I sold some of my land to people I liked—loggers and fishermen mostly, with children around the age of my kids, so they'd have other children to play with. Fred put in our road with an ax; at first, he and I and the Indian reservation were all that was there. His place is now a provincial park. He sold it to the government because he said he wanted Pender Harbour to have a public park on the water. He was born in England, and when he was fifteen he lied about his age and joined the Army. He was just a lad, still in his teens in the First World War, when the Canadians took Vimy Ridge from the Germans in a ferocious battle, in 1917. Thousand of Canadians were killed, and he lost a leg there. Doctors didn't know much about amputations and sealing off nerves then, and he's suffered terribly ever since. Fred wasn't very friendly with some of the locals, but he and I got along. He was like a passionate parent, the way he protected the wildlife on his place—deer, rabbits, raccoons, even wildflowers—and would threaten to shoot trespassers. I would often see him prowling with his gun at night, when the pain got to him—especially when there was a full moon. I used to go over and check to make sure he and Clara, his late wife, were alive, because nobody saw them for months at a time. We shared many a hot buttered rum, and I liked old Fred. I'll miss him."

Late that afternoon, we went through a pass to the north end of Calvert Island and anchored right below a dock whose most prominent feature was a telephone on a post.

"What a funny place for a telephone," I remarked, while we were eating an early supper on deck.

"Not at all," John said. "That's to call the Telephone People, who live on the hill above. We're going to visit them as soon as we finish supper. They run a big booster B.C. Telephone station for Namu, Bella Bella, and Ocean Falls. Their names are Andy and Nell Olsen, and they are the only people living year-round on Calvert Island, as far as I know."

As soon as we had eaten, we moved the *MoreKelp* over to the float, next to a motorboat that John said was the Olsens'. He went to the telephone and called them, and in a few minutes Andy Olsen appeared, driving down the hill in a jeep. He got out on the dock to greet us: a tall man with glasses and a short beard, wearing a white T-shirt and bluejeans stuffed into high boots. It was still daylight, bright and sunny, as we drove back up the one road, which led directly to the blue door of a freshly painted white garage. Beside it was a substantial house, white with robin's-egg-blue trim and a pitched red roof, and there was a garden full of bright flowers and vegetables, with a greenhouse. Andy's wife, Nell, came to the door to greet us: a good-looking woman with long, brown hair, wearing a yellow sweater. The Olsens were both so youthful that I was surprised when Nell told me later that they were grandparents. The house inside could have been a transplant from one of the more prosperous sections of Vancouver: wall-to-wall carpeting, a Franklin wood stove, big picture windows overlooking mile after mile in all directions of wilderness—Canadians always call it "bush"—and a channel where I could see two fishing boats with their poles up.

We had a cup of coffee, and then made a tour of the garden and the greenhouse. After that, Andy drove us up a boulder-strewn road that ended at a relay station, where a giant relay dish was mounted on a tubular metal frame on a high plateau—a treeless height of sheer rock, with occasional patches of low green scrub.

During Andy's detailed explanation of how a telephone relay system works, my attention kept straying to the dazzling view below. From this height, I could see exactly where we had fished, all the way out to Goose Island. Yet, on the rare occasions when I have had to call northern B.C. or Alaska, what I always see in my mind's eye while the number is ringing is that relay dish that I only glanced at then, with my call bouncing off it—I fancy my words as being wrapped in a clear plastic ball—from the top of Calvert Island.

Back in the house, Nell suddenly put a hand on my arm and her finger to her lips. We went over to the picture window and looked out, and at the very bottom edge of a long field saw dark shadows against the green trees behind. She handed me the binoculars, and when I looked through them the shadows materialized into a pack of wolves. "There are three especially that are quite tame," Nell said in a low voice. "We often feed them."

I handed the binoculars to John, who, once he had raised the glasses to his eyes, seemed unable to put them down. "God, what luck!" he said hoarsely. "I've been hearing about the wolves on Calvert for years, but I didn't think I'd ever see them."

Andy slipped out the door with a bag of food, and we watched him walk down to the edge of the field and throw the contents out. He turned to come back, and in a flash the food disappeared, and the wolf pack with it. He rejoined us, set up a projector and screen, and showed movies he had made of the wolf pack, close up and around him. "When I arrived here ten years ago," he said, "I was prepared to shoot every wolf on Calvert Island, but my opinion of them changed completely. The two brown ones and the black one you see in this picture follow my truck up and down that two-mile road strip and play together around it when I stop."

I asked Nell if she was afraid of the wolves.

"No," she said. "I enjoy them. We love being all alone on Calvert Island; sometimes we go weeks without seeing anybody else. We have a six-week vacation coming up and we want to stay right here. Where else could we go and find exactly what we want?"

The Telephone People drove us down to our boat and stayed to have a drink on deck, surrounded by a colorful array of our wash hanging up to dry. We spent the night at the dock, and slept in until six the next morning. It was a gray day: gray sky and gunmetal-gray water. On the way to the fishing grounds, we passed Jimmy Chambers's *Kitty D.* and three large charter boats for sportfishermen, circled by small satellite dinghies with outboard motors. The sportfishermen in their tiny boats were pitching in the waves, their lines in the water, apparently oblivious to the buffeting of rising winds. When we reached the open Sound, it was so rough that, while John fished steadily from the stern, I sat on the bench by the steering wheel with my feet braced against the compass shelf and had to hold on to the seat with both hands to keep from falling off. Eventually, John put down the stabilizers, and I went out and watched while he ran a little brown sail I had never seen before halfway up the mast. Once up, it didn't come down the rest of the summer. "Sails are very helpful in a following sea," he said when he came in for breakfast.

I was cutting the green mold off the last of our bread, so I could serve poached eggs on toast. "When we have several weeks of northwesterlies, some boats carry a bed sheet overhead," John continued. "About fifteen years ago, I had a dear friend, Gary Kulhane, now living in Ireland, paint me a lovely big fleecy sheep on my first boat sail. It had the words DON'T BE A, and then the huge sheep under it. It was a warning: I don't much like sheep. Twice every day, at slack tide, a fisherman has to decide where to go; or listen to everyone else's idea of where to go on the radio and go someplace else. A sheep follows the flock, and my point is that you can't do that; you have to think like a fish, not a sheep. I wish I still had that old sail. I don't remember what happened to it."

After a while, it began to rain hard. John had put on his oilskins and was fishing happily when I went to the door to see how he was. He looked up, the rain streaming over him, while the boat rolled and wallowed in the deep groundswell. "Thank God I'm out of sight of those bloody little boats," he said.

I wrote the word "groundswell" on my typewriter and looked at it. Swelling ground. Two words in perfect combination to describe what the surface of the sea was like now. When my typewriter fell on the floor, I gave up trying to work and baked cookies. Rough weather always brought on a compelling desire to cook. Some of my best meals were made when I was hanging on to the edge of the sink and grabbing for crockery and when canned goods were crashing to the floor. When I had finished using the oven, I opened the pilothouse door for fresh air and saw a great pile of springs in the fish box. Rainwater was dripping off the end of John's nose, and he came in shortly to leave his glasses on the sink. "I can't see anything with them in this rain," he said.

I turned on the Mickey Mouse radio and heard a man say, "I haven't had breakfast yet. Every time I start to cook, I find I'm heading for the rocks." He was interrupted by a voice that was becoming painfully familiar; it was that of a man who never stopped talking, was full of advice, and ended every sentence with the word "Okeydoke." When John came in to change his wet gloves for dry ones and heard the "Okeydoke," he said, "That man drives us *all* crazy because he talks about what fish are being caught and where, and that brings all the other fishermen around."

The rain stopped eventually, and eventually John came in, rubbing his cold hands, obviously delighted with himself. "Sixteen springs, six of them mild-cure," he said. "Time to stop. Peter is meeting us in the bay near the Telephone People, with a new generator for the *MoreKelp*. My old one has just about had it, and the spare is not much good."

We ran back into the bay, stopping to talk to Jimmy Chambers, who had taken a look at the weather and stayed in. We kept going, and straight ahead was the *Diane S.*, already anchored. John pulled up the poles on the starboard side and tied up to her, and Peter Spencer came over, carrying the new generator, followed by Yvonne, with two loaves of fresh brown bread and a pie covered with whipped cream and colored sprinkles. She had no idea what kind it was, so John said, "We'd better find out!" and cut it into four pieces and handed it out. The pie disappeared in a minute, we were all so hungry. Yvonne said, "We ate so fast you guys may not know what that pie was. It was pumpkin."

XXIII

We awoke the next morning to a thick fog; we couldn't see beyond our two boats, side by side. The fog seemed to be lifting, so everyone went out fishing, and we passed Cliff Gissing on the *Shaula*. His conversation on the weather conditions over the Mickey Mouse was accompanied by deep sighs of distress, as the fog rolled in again, more dense than ever.

"Are you there?" Cliff called. "I had to go out and look at my gear and lost track of you. You have two trollers beside you, and one behind you. I am almost blinded by the light, when I come outside after looking at the radar screen. My set has a range of about four miles, and if you stay within twenty I'll be able to pick you up later if the fog continues. I'll tell you one thing, John, despite all that's said about the loss of skill, there's a real built-in safety factor on these sets. Now I wouldn't go out to sea without one."

"If this fog continues, I'm going out to fifty fathoms and drift," John replied. "I can't see seaweed and I can't see birds, but I can still see fish boats. The way this damn thing has been up and down the coast, it may hang on all day." He went to check his lines, and when I happened to look out, he motioned me to join him. "Look at this!" he exclaimed, pointing to a huge halibut lying across two sections of the fish box. "I found that forty-pound halibut on the end of the pig line!" Just then, it jumped so far up that it almost hurled itself out of the boat. John thumped it hard, and it fell back with a soft flutter, as if the slightest breeze was passing through it, and then was quiet.

John was putting his gear away. A few minutes later, he went in and called Cliff over the Mickey Mouse. "I'm going into Welcome Harbour," he said. He pulled up the poles, and moving cautiously, we arrived at a harbor that was already filled with boats—trollers, sailboats, and big yachts. The largest boat in the harbor was an oversized vessel, the *Narvik*, a former yacht that was now a charter boat specializing in sportfishing. Its refrigeration plant was making a terrible racket, and its deck was stacked with dinghies. Small craft were running back and forth to the nearby beach, where we could see suitcases and duffel bags piled up on the shore, and, through the binoculars, middle-aged men in slacks and khaki sports jackets walking nervously around at the water's edge.

Surprisingly, because it was so foggy, we heard the drone of an airplane engine as we were anchoring, and watched an amphibious plane descend through the cloudy mist and land on the water, coming almost up to the beach. It disgorged a fresh batch of men into the dinghies, who exactly resembled those who were leaving. The new arrivals were taken to the beach with their luggage—which made a new pile—and the dinghies then picked up those who were waiting to go, with their suitcases, duffels, and cardboard boxes—which John said contained fish they had caught—took them out to the plane, and it departed with its homebound passengers.

I observed this whole operation with great interest from the pilothouse while John was outside cleaning fish. The plane ran along on the surface of the water, lifted, and disappeared through the fog. I could still hear the droning of its engine as I went back to John. He had finished his work and was pulling the dinghy down. He put it in the water and said, "Do you remember my telling you about Jonesy, my old pal who fishes steelhead with Al Perkis? We're going to row over and see him now."

As we approached a small troller, a large black Labrador dog barked at us from the gunwale. John stopped rowing for a moment and laughed. "That dog is twelve years old and he's famous for jumping off the boat when Jonesy goes into a harbor, swimming in, and meeting him on the beach."

I could see a stocky man with glasses who looked about fifty hastily putting on a plaid shirt, buttoning it as we came on board. His poles were still down, as if for fishing. "I don't dare pull up the poles, because my dog jumps overboard," he said after he had greeted us. He brought out some beer. "I've had a terrible time getting her back in when she does, because she's developing arthritis in her hind joints." He sighed. "I like to go to Alberta in the winter and hunt—mostly birds—and now she's getting too old for that, too. I've just come down from the Queen Charlotte Islands. I was staying near Masset, and there were seventy-four fishing boats there when killer whales got caught at low tide at the entrance of the harbor. The whales ran in under the floats and you should have *seen* them floats go up in the air! And all the boats with 'em!"

Jonesy began to fidget nervously, and when he told us then that he was invited for a roast-beef dinner on another boat we left, and rowed over to the beach, which was now empty. The newly arrived batch of sportfishermen were already out bobbing up and down in the dinghies, their fishing lines in the water, keeping close to the big charter boat, in the fog that enclosed us all.

We took off our shoes and walked on the beach—a long, sandy strip with small piles of driftwood—dipping our toes into the blue water that lapped at the edges. We picked a few blueberries in the bushes above, and when the sun broke briefly through the fog, we took a sunbath on the smooth sand. We heard a whooshing sound and sat up. John pointed to a large black whale that was spouting and rolling. "I think that whale is scratching itself on the rocks out there," John said. A chilly wind arose, and we put on our sweaters. A battalion of geese flew overhead in formation.

That evening, we had one of our best golden sunsets, but John looked at the misty clouds that hovered on the horizon glumly and complained about the fishing days he had lost. "I loved this day," I said, feeling like a traitor. "I don't wish you any more fog, but it was a delightful day. I felt the way I did once when I played hooky with my father. He was driving me to school on a beautiful sunny day, and suddenly he just swerved off and we went to our cabin in the country instead. He called his office when he got there and gave some excuse, and then we went horseback riding. I opened a can of Campbell's tomato soup for lunch, and then we took a hike and went home. It was one of the nicest days I've ever spent. Like this one."

When the fog fulfilled John's gloomiest predictions by reappearing in the morning, he announced that he was going to deliver his load of fish to Seafood Products in Port Hardy anyway. This meant crossing Queen Charlotte Sound through a traffic channel that was full of tugs with their tows; log barges; and large ferryboats—a hazardous journey in bad weather without radar. He hesitated, waiting all day for the fog to lift, and spent the time installing the new generator and putting a new roll of paper in the recorder-sounder, while I made up leaders and worked at my typewriter. By the following morning, when the fog continued, John had found that another troller, Jack Burroughs, on the *Rococo*, was leaving for Port Hardy. Talking over the Mickey Mouse, he and Burroughs decided to go across together, with the *Rococo* in the lead. Burroughs had loran, and therefore knew his location at all times. "He sets location buttons for where he wants to go and pushes a button that says 'Follow,' " John explained. "His loran tells him to turn right or left to keep on course, and buzzes when he gets to his destination; but he doesn't have radar, either, so he can't tell any more than I can who else is around or in the way in this fog." We started off in a rolling sea. The teapot fell out on the floor and slid halfway down the pilothouse before I caught it and laid it in the sink on a wet towel. John was standing with both hands on the steering wheel, peering out the window, tension written into every move he made.

"Why don't you get a radar set?" I asked.

John opened the compass window in front of him. "I think we can see better with the window open. Keep watching, will you?" he said. There was a long pause, and I wondered if he had even heard my question. Without moving his eyes from staring directly ahead, he said, "I think all us fishermen feel this deep urge to simplify. The mental worry of owning and upkeep on two houses, one floating—containing a mass of machinery and electronics—the other on land, seems like Mt. Everest. I *fear* extra equipment and gadgets. I'm *such* a messer. I've proved to myself that the more machinery I possess, the worse the mess is. I've had a lifelong desire for the simplest life, and an awful fear of losing any more of my so-called spare time."

Another pause. Then: "If I were in the habit of running fish into Prince Rupert or Vancouver now, I'd be crazy to be without radar, in view of the enormous traffic and the danger of getting between log barges and their tug. But if you stay in areas where there is less fog, I question whether radar is worth it—especially in poor production years. You have to earn more with it; you have to pay a lot more insurance premium on the greater amount it adds to the value of the boat. I don't do what I'm doing now often. You should see the look, though, that comes over the faces at the fish companies when someone like me comes in *with* fish but *without* radar. It drives them absolutely nuts."

So far, we had kept the *Rococo* in sight. Without warning, the fog thickened, so we could not see anything in any direction. John picked up an old-fashioned long tin foghorn that he had placed on the seat beside him before we started, leaned far out the window, and blew. He continued to blow fervently on the foghorn, and we plowed ahead blindly. Then I blew the horn, and he just steered. There was no telling what was on either side of us, behind us, or ahead of us, but I was counting on John's highly developed sixth sense in emergencies. It was the most nerve-racking journey we ever made.

When we were safely across on the other side, John put the foghorn down with a great sigh and said, "That was a hairy trip. Do you realize how dangerous it would have been if a tug had come between us and Jack?"

Yes, I did.

The fog perversely lifted as we moved into quiet waters. I went outside to see where we were. Behind me, I heard someone talking over the Mickey Mouse and John replying, but I couldn't catch their words. When I came back in, John was sitting quietly, staring out the window. He looked extremely pale.

"Are you all right?" I said anxiously. "What's wrong? Who was on the radio?"

He took one hand off the wheel, ran it through his hair, glanced at me and out the window again. "That was Jack Burroughs, on the *Rococo*," he said. "You can see him now ahead of us. He was telling me that while we were all crossing the Sound he looked behind just in time to see a large tug come out of the fog, right between us. It came so suddenly there was no time to warn me; it was already there. It probably saw the radar reflector on my rigging; anyone's radar screen can pick up that blip." We looked at each other. "We were damned lucky," John said. "Just lucky."

We delivered our fish at Port Hardy the next day, and John turned around and went out again that same afternoon, to avoid a flock of gillnet boats coming in from an opening; they had been allowed three days this time by the government to fish. "Those closures, which are correct to enhance the fish stock, are a conservation method that applies only to commercial fishermen but not to the two hundred and fifty to three hundred thousand sportsmen who are fishing," John said. "It just doesn't make sense."

I barely had time to do a laundry while he got the groceries and ice before we were on our way out, passing one gillnetter after another coming in, running low in the water, heavy with fish. Reg Payne was at the cannery at the same time we were, and we agreed to a rendezvous that evening at Port Alexander. We invited him to have dinner with us there, and on our way John told me that before Reg became a fisherman he was a tugboat captain and in charge of pleasure yachts remodeled for war use during the Second World War. Because of his skill as a navigator and his knowledge of the whole coast, he was then lent to the Americans in Alaska.

Reg was already anchored and waiting on his handsome troller, the *Saturnina,* when we arrived at Port Alexander. A much larger vessel than ours, it had been designed especially for him after a North Scotland blueprint. He fished halibut every spring, and had a special gurdy on the starboard side for hauling in the big fish; he also carried a spare trolling pole in a slot in the bow that always made me think of a male narwhal, which has a long, narrow tusk that extends directly out the front from its lower jaw. Everything—even the rope and chain for his elaborate anchor winch—had its own storage slot. I could appreciate the details of his boat better now than when we had encountered him at the beginning of the summer, and I showed off my added knowledge by asking him if he was sleeping well in the foc's'le, to demonstrate that I knew better than to refer to the area belowdeck where he lived as being "downstairs."

Sitting out on the deck, we finished eating Chris Sondrup's delicious smoked coho, presented to us as we were leaving Port Hardy, and plunged into the main course, a cottage ham baked with pineapple; boiled potatoes; and fresh vegetables. The two men began reminiscing. They talked about a fisherman they had known as Eric. "This one time, Eric took his brother fishing with him, against the wishes of the family, because he was a psychotic who had been in an institution," John said. "Nobody heard from them, and up on the grounds at Hecate Strait, everyone noticed that his boat, the *Comfort II,* stayed where it was anchored on the Ole Spot when the others took off to fish. They wondered, but sometimes people do rest for a day that way. The next day, they called him on the phone and nobody answered, so finally, that evening, when they saw his boat just rocking in the water and nobody coming out ever, one of the other fishermen went over in his boat and jumped on board. He found Eric slashed to pieces, with blood everywhere, and the brother had disappeared. He had obviously jumped overboard; his body was washed up several weeks later."

Reg said, "I knew another fellow—a bachelor, also named Eric—but this one lived in Port Hardy and was known as Silent Eric because he never talked to anybody on the radio—who was found late in the summer on an island in Rivers Inlet, out of his mind. They thought he had been there about three weeks without any food, and they found his boat sunk, so they figured he had run it onto the rocks. They took him to the hospital at Bella Bella, where he stayed two months, and then the hospital authorities said he was well enough to go home. So he left for Port Hardy, and was never seen again. Nobody knew he was missing because the hospital never notified anyone at Port Hardy. Much later, fragments of his boat were found on Goose Island. No one really knew what had happened, except that the weather had been terrible. It's very important, especially for people fishing alone, to keep in touch with one another. When something goes wrong and you don't hear from them, then you can investigate."

"One of the reasons I've always been fascinated by meteorology, I guess, is that it's *survival,*" John said, taking our plates away, while I brought out a crisp I had baked with rhubarb fresh from Lillian O'Connor's garden. "We still have to be fairly good amateur mechanics and electricians, but the older fishermen before us did major shop jobs at sea in gales! Now we do minor things and howl on the bloody phone and the Coast Guard picks us up, but fifty years ago, this halibut boat broke its propeller shaft on a four-to-five-day run from Alaska to Prince Rupert. It had a small sail like mine, so the fishermen aboard made more sails out of the blankets and got some movement from that. Then they cut part of the boom into sections and used it for an axle, and rigged up a rope drive for it from the winch used to pull up halibut gear. They made two primitive side wheels—really big crosses—from the pen boards in the hold, and ran one on each side of the boat, with the boom sections as the shaft across the middle; and they made some kind of wooden bearings and had a cup of grease to keep them lubricated. They had good weather and made one or two knots an hour almost to Dixon Entrance, the ocean passageway to Canada, and were in sight of land when they ran out of fuel. They were out of grease and almost out of water, too, but they still had halibut to eat when another boat spotted them and towed them to shelter." He shook his head. "When you think of it, that could happen to any of us if our batteries went flat."

The rhubarb crisp was a great success, followed by hot buttered rum. "I shall always remember the sixty-five-year-old Catholic priest who had a new twenty-year-old curate with him on his boat, and they were in their first winter gale off Bamfield, on the West Coast," John said, as Reg was leaving, holding on to the rigging, with one foot on the gunwale of our boat, the other poised on the rail of the *Saturnina*. "The little twenty-year-old bugger got down on his knees and began to pray and the old priest said, 'You can pray all you like later. Right now, grab that pump and don't stop pumping.' "

XXIV

It was getting toward fall and the weather was increasingly unpredictable. There were fewer fishermen on the grounds, as those with children in school headed for home. For several days after we said goodbye to Reg, we had wonderful fishing weather; the hold was filling rapidly with salmon. I had developed a regime of work for myself, too. This particular morning, I had awakened and was in my bunk, thinking out my program for the day, when I felt a rush of damp air through the opening above me, where I had pushed the hatch cover back during the night. It was five-thirty—late for us to start our day—and the sun had already risen, showing itself in a few shots of lemon yellow streaking across the horizon in a light-gray sky. We were trolling off a formidable rock that rises abruptly out of the water. It was called Blenheim Island. A scattering of low satellite rocks on either side, John told me, required the most careful navigation—a sort of horseshoe-shaped maneuver for us and for the three other trollers who were also there: a large number of fishing vessels for an area with such limited boundaries. We were all moving very carefully.

I finished my housework: the breakfast dishes washed and put away, the toast crumbs brushed into a dustpan held to the edge of the stove, and a light sweep of the floor with the broom. I rubbed three brass-and-copper lures with a rag and Brasso until they shone so brightly that I did not think any fish could resist their appeal on John's lines. I visited the cockpit to admire two small springs and six large cohos that John had caught on the early-morning slack tide, and gazed with some guilt into the sad pop-eyes of a lingcod, which occupied—and filled—one whole fish box. Its outsized head, with its great square jaw, hung over the box, dwarfing the long, graceful body, with its tapering tail. I remembered that the first time I saw one of these greenish-brown fish, with its monstrous head, I was repelled. Now it looked beautiful to me. I sat down beside the fish box, leaned over, and touched the cold wet body of the lingcod. Just then, John brought in another fish, a large spring. Thump thump. Death in the Cockpit. But I no longer minded so much.

"I have accepted your philosophy," I said to John. "I have accepted my place in the food chain to which every living creature belongs. Here, on this boat, I see it in bold outline."

John looked up and laughed. "I'm glad," he said. "That makes it easier for both of us."

The sun had come out, a bit dimly, after all. I retreated to the pilothouse to replan my day. A given day. Why work? I had a novel I wanted to finish reading, some letters to write, and I wanted to fill in all the red parts of the needlepoint glasses case I was sewing. If the sun stayed out, why not a little nap on the deck, too?

First, the novel. I went down to get it in my bunk, had it in my hand, and was climbing up to the pilothouse floor when I saw John's feet on a level with my head. "Steer!" he shouted down at me. "My God. Don't you see the fog?"

What a shock! In that brief moment when I was down below getting my book, a damp shroud had surrounded us, closed us in a cocoon of only a few feet of space, on a billowing sea of menacing gray water on all sides. Where were the three other vessels? Where was the coast? Blenheim Island? The dangerous low rock piles? All had vanished, and we were alone in a tiny, chilling world, moving in this aloneness through a deadly veil that parted to let us pass through but would not let us go.

John went on fishing, and I steered, with the depth sounder ticking away at my side, flashing orange-dot signals around its face. John put his head in at the door. "Keep to a south-southwest course not less than fifty fathoms," he directed. He came back to say that if the sounder showed less than fifty fathoms I was to turn south—or was it west? I immediately could not remember which.

I fell into my familiar behavior pattern in fog, moving my head continuously in a half circle while I steered, so that I could see any boat coming toward us or moving beside us—a half circle that included a quick glance at the sounder each time.

Living with fog: always the same. First, the excitement of the danger. Then minutes become hours, and nothing happens; no boats or other threatening objects appear outside the pilothouse windows, beyond the point of the bow. Nothing to see but more fog and gray water, with an occasional piece of wood bobbing by, or maybe a small bird, diving bottoms-up into the lazy rippling waves. Boredom comes first; then sleepiness, which must be fought off with the nervous search for anything of interest on the compass shelf; followed by an overwhelming urge to eat.

I passed through all three stages. I considered the possibility of slipping from the seat and walking backwards to butter a piece of Britl-Tak, when I glanced at the sounder. Tick tick. *Forty fathoms!* A cold chill started at the back of my neck and traveled down my spine. Had John said to turn west to find deeper water? Or was it south? Tick tick. *Thirty fathoms!*

I turned the wheel to the port side. Nothing happened. I tried to turn south, reading the compass. The sounder still read 30–32–30. I dashed to the pilothouse door. "Left or right?" I shouted.

"Right!" John shouted back.

But, as I turned the boat, the sounder was already reading fifty. I fell back on the seat, almost sick with relief.

I was still steering a half hour later when John came in, made a pot of fresh tea, and took over. "I just went over my lines. We have two more cohos," he said.

I had already opened the window and John, standing by the wheel, had his head half out, straining to see ahead. I hugged my mug of tea, looking into the fog. I saw John suddenly move the throttle ahead and double our speed. I quickly glanced at the sounder. Tick tick. *Twenty fathoms!*

Staring ahead, I could see nothing to cause this sudden change of speed, so I knew we were crossing a shallow spot he hadn't expected. I was stunned. It was so completely unlike John, whose mind read the bottom as if he could see it. Always. We picked up speed again. Tick tick tick. *Nine—eight—four!*

John moved the throttle handle forward as far as it would go. I felt as if we were flying. I glanced back through the pilothouse door. The lines were flapping in the air, loose and wild, and the gear was bouncing. A pig was lying on its side at a very strange angle. "I think you've lost a lead," I said timidly.

"A lead!" John said grimly. "We've lost four of them! We've also lost all the bloody gear by now. I can't get away from the rocks."

I stared at the sounder. Tick tick tick. *Six—eight—twelve.* Another five minutes and we were back at a fourteen-fathom depth. The sounder jumped to thirty; finally, to a comfortable fifty. Too late for the fishing gear, but we had not gone on the rocks, and we were in deep water again.

I moved aside while John pulled down a chart from the rack above and looked at it. He turned the boat around, in the opposite direction. "You steer," he said. "I've got to look at my gear. It'll take me at least an hour to put it together again, and I'll be exhausted. That's it for the day. We'll go to Goose Island to anchor."

It was only then that I realized that the fog had lifted.

XXV

We had lost so many cannonballs that John cut our trip short and we returned to Port Hardy. When we went into the marine-supply store at Seafood Products, John bought the last three they had. We were still one short, but so many fishermen had lost leads during this foggy episode that we were glad to get any. While John was delivering his fish, I went to the public telephone to make a long-distance call. The line of fishermen waiting to call was so long that I went upstairs to the office of Seafood Products to use their phone. My call took only a few minutes, and as I was about to go downstairs, Don Cruickshank, one of the owners and the manager of Seafood, invited me in for a chat. I had told him a while back that someday I would like to talk to him about his work.

Don Cruickshank said that some fishermen had left, so the pressure had lifted a bit. What he wanted to tell me was how John had helped when he and his partners were getting started at the cannery—something I knew nothing about. I only knew that at the end of every trip, after John unloaded his fish, he climbed the stairs to the cannery office and collected slips for what he had sold. Occasionally, since he carried very little cash on the boat and most of that in quarters in a jar in his "office" under the mattress of his bunk, he would come away with some cash to buy groceries or marine supplies, or for a restaurant meal. I never saw the checks for the main fish payments.

Don Cruickshank was a very quiet, direct, no-nonsense young man who had told me at our previous talk: "Some people enjoy painting pictures. My only hobby is my work. For me, there is nothing more beautiful than the clunk, clunk, clunk of a fish-canning machine." He was a devoted family man, so I had asked if he could think of anything besides fish and his family that interested him. He shook his head, and added, "Every clunk is a can of fish going through. That's pure joy for me. There are different strokes for different folks. I enjoy seeing things run."

He told me that he entered the fishing business in 1954, right out of high school, working at a marine-supply store for fishermen in Vancouver in the winters, and in the summers at float camps and in the salteries that existed then up and down the coast of B.C., where he hung and repaired nets. Eventually, the company he worked for made him manager of their central fishing area, and one of the customers for the fish he bought was Seafood Products in Vancouver, owned by a man named Peter Van Snellenberg—the third generation of his family in the fish business.

"The outfit I was working for went broke, and I was talking with one of the men I had worked with, Norm Manson, about having a fish-processing plant somewhere close to the fishing grounds. Van Snellenberg wanted to secure a supply of salmon for Seafood and said he would put up the money for a plant if we would build and run it," Don went on. "We looked around and chose Port Hardy, because it's on the end of the road on Vancouver Island, and the closest spot to the Central Area fishing grounds—readily available to virtually every sockeye fishery. In this industry, by the way, salmon is king, and the others are handled to pay the light bill. Basically, B.C. Packers is the main monopoly; with us, you're looking at maybe between three and four percent of the total fish production of B.C."

Don went to the door and asked his secretary to bring us two cups of coffee, and continued, "We came here in 1966, and in 1976 four of us who worked here, including Ross Kondo, bought the company from Van Snellenberg. When fishermen sell their fish to us, some, who have difficulty looking after their own paperwork, have their payments booked, as we say, or credited to their account. They take out enough dollars for grocery money, and pick up the rest at the end of the year. In all the years I've known John, he never booked his sales but took a combination of cash and a check, which he had sent to the bank." He leaned across the desk. "In the middle of John's normal settling up in our first season as owners, he asked me if we had any cash problems. John fully understood the problems that relate to cash flow. You might have ten million dollars tied up in inventory, but you have to have cash, so you borrow that; and the more successful you are, the more fish you are buying, and the more quickly you can overstep your line of credit."

The coffee arrived, and he took a sip. "We were financed to the hilt then, and had a real fear of spending all our dollars before the end of the season. That particular day, John asked me what benefit would it be for him to book his fish. Well, buying raw fish is by far our largest cost—maybe eighty percent. You can temporarily hold off paying oil companies or can suppliers or fishermen's gear—all normal accounts payable—but *not* fishermen. The fish business is unstable, with a history of failing companies, so a successful buyer *must* retain the confidence of fishermen; you have to maintain that trust, that you are stable enough to pay them. On that particular payment slip, John needed the cash, but he said, 'On my next trip, remind me and I'll start booking my fish.' When he came in, we settled from force of habit in the usual way, and he left the office. He told me he was flying to Vancouver that night, and it so happened my wife was going, too. When I got to the airport, there was John, running toward me, waving the check I had given him. 'Don, we forgot to book the fish,' he said, and gave me my check back." Don picked up his coffee cup and drained it. "I expected him to collect at the end of the season, and I recall that we owed him a pretty significant amount, but he said, 'Oh, no, send me two thousand from time to time, when I need it,' knowing that by early spring we would have the cash. This is the type of gesture I appreciate from any fisherman, because it shows concern for our business, but I was even more impressed, coming from John, because he had the reputation for being a strong socialist. What he was saying to me was 'Don, I sure hope you make it.' He understood the need for capital in a capitalistic society, and was contributing his share of the capital to the success of our company."

Don said he would like to show me around the plant, and on the way downstairs he said, "I suppose you know that John is an exceptionally good troller. Any man who can troll and catch spring salmon knows far more than the average fisherman. With fishermen, I'm more often their source for answering questions, but with John it's the other way around. More often, I'm asking him questions. Furthermore, when he brings his fish in, you can't tell his first-caught salmon from his last. They all look as if they've been caught this morning. Fishermen are all different. Some are extra-conscious of their product, and others just sell fish."

We went on a quick tour of the plant, starting with the ice-making machine outside, and into the cannery building, where Don stopped, enthralled, beside a monstrous-looking orange block of machinery, which he called by its trade name, the Iron Chink. As it clunked, it butchered the fish, scraped out its entrails, and partially cleaned out the belly cavity, doing the work, he said with awe, of sixty people. To the tune of this clunk, plus the clink and clank of empty cans being fed into the canning machine and the thump thump thump of the filled cans being machine-closed, he guided me through a maze of washing, weighing, filling, canning, and cooking machines, surrounded by little clusters of human help: women, chiefly, sorting and arranging, in rubber aprons, boots, and gloves, their hair, which by law must be "contained," tied back with scarves and ties—splashes of bright color.

We went from that building to the loading dock, where youths driving forklifts with huge wooden boxes, or on foot, pushing fish bins on dollies, were ducking around us. Don halted at the foot of the office stairs. "One thing more you should know," he said. "Fishermen like John can catch fish without any of the latest electronics, but the new ones can't. The new fishermen know where they are because of loran. John will have landmarks lined up and readings marked on sounder paper to indicate where he was or the edge of banks. What he knows involved many many years of a learning process when he was younger. There are only a few fishermen left like John; they don't fish with the idea of fishing the boat but to outsmart the big salmon—specifically, the springs. You know when John comes in that it's going to be spring salmon, no matter what time of year—which is not what you get if you are fishing for whatever it is that makes dollars."

I stood back to let a forklift pass between us into the plant. Don said, "As for us, in almost any other industry—let's say, one that makes can openers—you know what the product will be and where the sales are. Here we produce what we can get in any given year, with no regard to sales, and it varies from season to season. We turn out canned salmon whose quantity is governed by God only knows what factors, and then scramble to sell it."

XXVI

The season was ending. We made one more trip out of Port Hardy, had a grand farewell evening with the Spencers, the O'Connors, and Chris at a local steak house, and Dave O'Connor drove us back to the boat.

We left the next morning with an enormous rhubarb root on our deck which John had pushed out of Lillian's garden in a wheelbarrow and somehow gotten back to the boat. (It was in the way of everything as we headed home, but it still grows green and lush every spring in my garden, producing splendid rhubarb.) John said, "In spite of all that fog and the strike, we had an excellent fishing season. We'll go up for a day or two and fish toward the end of Vancouver Island, and then we'll start back, stop at the house to unload this rhubarb, and go right on to Vancouver." We both went out on deck and waved to Lillian. As early as it was in the morning, a few minutes after six, she was at the window, signaling with a flutter of white cloth when we passed. "I have a lot of people waiting in Vancouver for fish," John said. "I wouldn't want to disappoint them."

With a new four-ton ice supply in our hold, we started fishing again. We fished all that day, and toward evening, right after slack tide, John turned into a lovely harbor that reminded me of Boom Bay. We were the only people there, but there was a broken old white boat up on the shore, and what looked like a barrel with two big flywheels, one on each side, which John said was an old sawmill engine. Logs and lumber littered the beach. We had just finished supper and were sitting on deck, John in his chair, I lying on top of the hatch, looking at the sky, enjoying a spectacular red sunset. I could see waves pressing through the harbor entrance, carrying clumps of kelp on their whitecaps.

"What is it that makes me think of Boom Bay here?" I asked.

He had made us each a hot buttered rum to celebrate the start of our last trip of the season, and he raised his mug and took a long drink. "Ghosts," he replied. "Because of old pals of mine who are gone. All dead now. We are alone with their ghosts. You feel their presence, and so do I. We're in Cascade Harbour. There was a lot of activity here in 1938: two stores, one float, one float house ashore, one sawmill, and two fish buyers, and not a soul more. Now there's naught here but mink and seagulls. We'll start with Brown. Brackish Brown. He was very robust, with huge bushy eyebrows, and a lot of hair in his nose and ears, who lived on straight starch—porridge and potatoes—all winter, from October to March. He saw very few people, and once a month he would row over to the post office in Shushartie Bay, about ten miles away, and get his mail. He grew a garden to attract deer and shoot them from his bedroom window, and he ran a very ramshackle sawmill, sawing up small amounts of lumber with great effort, living in a house with only the barest essentials."

"Was there a Mrs. Brackish Brown?" I asked.

"He was a bachelor, except for five or six months after the First World War, when he got a woman to come out from England by putting an ad in the paper. They corresponded, and he told her about his beautiful situation in Cascade Harbour. She came, and I suppose they married, but apparently his bride decided after four or five months that she *had* to find some way of getting back to England, and relatives promised her enough money for the fare back. Brown was very much against this, so she had to hide and run down a very rough, wet trail to the edge of Bates Pass. A man there, not a friend, saw her and rowed her to the post office at Shushartie Bay, across Goletas Channel. She was able to catch the Union Steamship there for Vancouver, and went back to England. Brown never forgave that man, and became his mortal enemy."

John shifted in his chair. "Now the Cholbergs," he continued. "There was Chris Cholberg, his wife, and a son, living on a scow that was beached in this harbor in the corner to the left there. It was partly sunk, so they lived for years in a float house that was set at an angle, and they bought fish for the Canadian Fish Company in Vancouver. Cholberg had a fifty-foot packer he called *The Silver Horde*, which his son, Irwin, used to run as far as Vancouver from the time he was fourteen. He paused. "There's a seal over there," he said, pointing toward shore, but I missed it. "Cholberg gradually slipped behind until he owed the company five or six thousand dollars—a fortune in those days—and as he made a hobby of studying law, he suspected the company might seize his boat as his only asset. So he turned around and got Irwin to put in a claim for back wages for running the packer, in the sum of seven thousand dollars. It wiped out the debt, so it all came out in the wash. Cholberg Senior had been foreman at a big Victoria shipyard in his younger days, and was a clever man, with a very keen sense of humor. The last time I saw Cholberg was in 1944. He'd row over to Shushartie Bay and get your mail, and he'd hand it to you with a glass of gin and want you to come over to the house and visit. And then there was Jones. Oliver Jones. Both Jones and Cholberg bought fish, Jones from 1934 to 1951. If you sold to one and not the other, the other wouldn't speak to you, but Jones had much the best-stocked store. He built up a remarkable wool business with Americans going back and forth to Alaska, mostly fishermen, who always stopped here. Mrs. Jones was a darned good buyer, and they had beautiful Hudson Bay blankets, pure gray wool mackinaws that were very warm, and Stanfield's underwear. When the Americans were moving back down from Alaska to Seattle, they really stocked up."

John hugged his mug with both hands and began to laugh. "Cholberg, Brown, and Jones had field glasses, and every time you went to visit, they weren't so interested in you as what you had found out about the other two," John said. "Sometimes two of them weren't speaking to the third, or all were not speaking at all. They would be watching every move the others made through their field glasses."

He leaned forward in his chair, squinting. "There's that seal now," he said.

"Where? Where?" I exclaimed.

"You missed it again," he said.

"Brackish Brown," he murmured, and shook his head. "He had a huge one-cylinder engine that drove his sawmill." John handed me the binoculars. "Take a good look. The sawmill's lying there on the beach—what's left of it. It was in very poor condition even then, and he used to crank and crank and shout, 'I don't know what's the matter today, you son of a bitch. I've given you the best gas, you've got water and a spark a mile long. Now, go!' Then he'd crank and curse, and if that didn't produce results, with the poor customer standing there waiting to get lumber, he'd say, 'That's enough for today. Let's go in and make tea. That s.o.b. may be in a better mood later on.' So you'd troop into the house, and he'd make this coal-black strong tea. He also used his body as a human belt-tightener to take up the slack on his drive belt when he was planing wood. He lived alone so long in the pouring rain and howling gales that many inanimate things became alive, with annoying human characteristics."

"What happened to all those people?" I asked.

"Mrs. Cholberg died first," John replied. "Then Mrs. Jones died. Jones wanted to go back to Australia or New Zealand—wherever he came from—but he never made it. He built a sailboat and called it the *Dingo*. He got one or two days off the West Coast, turned back, and sold it." He sighed. "Many's the time I've battled after a night of the Joneses' homebrew beer and a wee bit of moonshine, too."

"And Brown?"

"Brown stayed here for years, and used to write me glorious letters," John said. "He believed that Cascade Harbour was the potential center of the northland. Brown used to say, 'I can't see making Port Hardy so important.' It was wonderful to see such faith in this little place as Brown had. Why he ever thought it could be, I can't imagine, with such a surge back and forth of waves whenever there's a big blow."

He sat watching the waves thundering and churning at the entrance of the harbor, then said, "Jones sold his store to a company that later became part of the Prince Rupert Fishermen's Co-op, and Brown watched the store buildings as a janitor for the Co-op. Once, he lost a jack, and he shot someone who he thought had stolen it, and the police came, but nothing happened, so I guess the wound wasn't serious. Brown also built the most terrible boats. But they didn't leak. He was in the forestry unit in the First World War, cutting trench timbers. He told me that an officer he described as a real Army type—all patriotism and glory—came along and made a big talk about going to the front and said, 'All who want to replace brave men who have been killed, step forward.' Everyone stepped forward except Brown, and the officer asked him why he didn't. Brown said, 'Sir, I came here to teach sawmilling. I'm supposed to be sawing trench props, and I trained for that in B.C., and that's what I'm going to do.' And he did." John stood up. "There's the seal now, right at the entrance."

By the time I looked through the binoculars, it was gone again.

"He saw us and camouflaged himself among the kelp," John said.

John drank what was left of his rum. "The last time I saw Brown, I came when he was clearing out a stump with long roots that was floating around. He was in a rowboat, rowing and pushing the stump from his stern, trying to get it out of the harbor. He would shake his fist and lose ground, and say, 'You s.o.b., every time I start a boat, you're back, and I'm going to take you out in the channel and get rid of you.' He almost rowed into me, and he said, 'Oh, that's you, John.' I hadn't seen him for three years, but he went right on with what he was saying to the stump, and away he went, a really vigorous guy. Then bump bump bump, and he stopped twice after he passed me to shake his fist at the stump. He was gone for a good half hour, and when he came back I went up for tea and brought him back to the boat for a big supper."

John stood up, folded his chair, and stowed it away. It was getting dark, and there was a sharp chill in the air.

Early the next morning, John turned the bow of the *MoreKelp* south instead of north, and I knew we were heading for home.

XXVII

We took it easy on our journey home, with John stopping now and then to fish and with me indulging what had become my favorite pastime: reading backwards in John's logbooks. I was now reading the very first one—a dog-eared eight-by-ten, black-cloth book with worn red-leather edges whose first entry was dated January 27, 1935. John was twenty-three years old and, a few months before, had just bought his first boat, the *Snark*.

It was more of a journal than any of the other logbooks were, and reading it was a strange experience for me. When my mother was eighty, I asked her how old she felt inside herself. She immediately replied, "Eighteen," and I kept remembering that. Here I sat, beside the person I was reading about, but I was moving back in time to meet that exuberant youth I knew was still inside John: a partial stranger to me. There were familiar elements of the older John in this raw young stranger: the same high spirits; his delight in funny anecdotes; his penchant for worry; his joy in his friends and surroundings; his sensitivity; and a certain austerity, too. He wrote a lot about what he ate, and he listed his daily fish catch, in an offhand manner; but what came out most clearly was his total love of the life he had already chosen with such deliberation.

I usually read when I was sitting on the deck outside on the other side of the fish box from John, holding down the flapping pages with my hand, glancing up from time to time at the man I knew so well—with his neat white mustache and close-clipped beard—who did not seem at all old to me, because I had never known him when he was young. Now he usually wore a hunting-type cap with a visor. It had flaps, pulled halfway down over his ears because of the cold fall weather, and was red, orange, and black plaid on a white wool background. It made him look like a country squire—at least, around his head. The rest of him was suitably clothed in either his oilskins or his red-and-black wool mackinaw, and his brown wool pants. I had seen plenty of pictures of John as a young man around our house: the long, clean-shaven chin; black hair combed severely down or in the crew-cut style of the time; lean, as he still carefully was. He was usually photographed either on a boat or with a boat in the background.

The journal, written first in blue ink that ran, and later in black, waterproof ink, in a clear handwriting that within a year had deteriorated into the almost indecipherable scratching with which I was familiar, began at Coal Harbour, in Vancouver, on January 27, 1935, at 11 a.m. The following day was labeled HOME, which meant Cherry Point, at Cowichan Bay, where he had grown up. The next entry, in March, described a "gentle breeze" that accompanied him en route again to Vancouver. "My wake stretches out bubbling white for many yards astern—the only disturbance in an otherwise unruffled calm," he wrote. "Spring, I hope, is here. Conflicting thoughts: starting with gasoline fires, sunken 'deadheads,' engine trouble, etc., flit through the idle brain. The super-clearness of the day makes far-off objects deceptively closer than they really are."

The following day, he explained that the "muck on this page"—a substantial water blot—had been from "a large sea breaking on the bow suddenly, causing me to shut this log with a bang and attend strictly to business," until "finally, in a blaze of glory, the *MV. Snark* nosed its way into Vancouver Harbour at 12 noon."

He wrote about whitecaps and a "squamish gale," fierce gusts of cold winter wind blowing off the glaciers of the Coast Range, "that bane of our existence, the thorn in the side of all those who travel in Howe Sound." He described how he "took a few seas over the bow and rolled like nothing on earth while turning past the Point." On the lower half of an early page, whose upper half had been washed out by seawater and had "rough Sea, Windy weather & a rocky coast" scrawled across it, he wrote down an anecdote he had heard: "A certain Mr. Bill Cates, a tug captain of repute, learned to play tunes and bugle calls on his tug whistle. One day, while proceeding up the Fraser River, he passed another fellow with a tow and disappeared around the corner. The other fellow soon heard a faint whistling, which later materialized, after the third or fourth repetition, into an unmistakable

'How-dry-I-am' (Ta-ta-ta-ti), and Mr. Cates was discovered high and very dry upon a sandbar."

The young John talked of smoking cigarettes (firmly banned from our house). He enthusiastically told about large breakfasts of three eggs, six rashers of bacon, with hotcakes to top it off. Skiing was important, and he already pursued it in every free moment. Each day, he faithfully wrote down the number and variety of fish he had caught—mostly salmon, and an occasional halibut. He spoke of "the idle rich in their various crafts from sail to steam . . . on the way in, this being a Sabbath evening," and he attempted to compose "a little homily upon the 10 fishing commandments," which he unfortunately did not include.

He used his boat for other purposes. During the first fishing strike he experienced, in May of 1935, he delivered mail and food to people who "were all running around short of grub due to the strike," commenting on "the hot sunshine and a sea breeze all day long, and beautiful starlit nights— What a life!"

The first time he went through the Yaculta Rapids, the gateway to central and northern fishing grounds, where he subsequently spent most of his fishing life, he was forced back because of the tide and had to wait five hours before he could make another, successful attempt. Afterwards, he wrote that "it was so warm we could take off our shirts and lie on the stern, steering with our feet," and continued, "Islands, islands everywhere, and pleasure boats appearing from many of them, hurrying home to the call of office work. Thank God, no more of that for me."

Fishing in the north, he mentioned strong tides that "hum past our anchor ropes." He periodically noted down the many hours spent molding hundreds of pounds of lead into bars for ballast. On a separate page at the back, he carefully made a list of equipment needed and the lead-ballast recipe, with a warning to KEEP ALL WATER AWAY during the molding process and to "open the molds with a hatchet blade & not hammer them, as it may spring them."

The first time he fished off the West Coast, he wrote, "Quite an experience fishing on the broad Pacific." He noted that he caught twenty-two-pound and twenty-five-pound springs and a coho before a strong east wind and sheets of rain made visibility "nil"; and that he went into a harbor and "am now about to sleep." A month later, in July of 1935, he arose at two-forty-five in the morning to cross the Nahwitti Bar near Cape Sutil at slack tide. "Whilst sitting in the pilothouse meditating, I suddenly beheld, much to my consternation, the port bow pole bend in an arc to the water, wiggle violently, and finally crack in the middle with such a report that it made a fisherman 300 feet away come out of his cabin. Another of these halibut, judging by the state of the hook, and the flesh thereon."

He described "thousands of dipchicks" in the water around him, and days spent fishing around Goose Island and the Goslings; and also many nights at Cascade Harbour. "Enormous swell and bitter westerly, almost gale force outside Cascade, with a nasty back slop off the rocks," he wrote, and he said that the clothes he was wearing, "2 pairs of pants and 3 sweaters, felt like nothing, and this is July!" Soon after that, he added, "Motto for today. Don't read a book while dragging 20 fathoms in Bates Pass, however far you *appear* to be off shore!!," and he drew a rough cross, with the inscription beside it, HERE LIES 30 LBS. LEAD & 6 SPOONS.

In August of that first year, he reported: "At Namu Cannery. Arose 8 A.M. and had a shower bath, after 2 months. What a joy! The harbour was black." Several days later, he caught a "52 lb. Spring: $44.16 by gum! & 14 Coho."

He experimented with different varieties of lures, which he described in detail. He noted, later in the summer, he had had "no time for five days to keep the log . . . fishing 14 hours a day and more"—a pace that became normal in the years that followed.

He had an exciting September in 1935. He anchored with friends in a little bay, and in the middle of the night it blew so hard that "we had three anchors out and three ropes ashore! And still it blew! No sleep tonight. Had to put out two more anchors—making 5 in all—and scramble up a cliff, this being low slack, and untie the port shore line and swing around facing into it. Then had to take another anchor in the skiff and dump it. Sheets of rain and howling wind. The other three boats had a mulligan stew of anchor lines and shore lines when they went swinging. Started up the engine and ran it for about 15 minutes, but the wind slackened a little. What a night! This is midnight and we are still hanging on by our teeth."

He hit a log "full smack but apparently no damage done," saw the smoke from forest fires on the way home, and arrived in Vancouver on October 3, 1935, at two-thirty in the afternoon, stopping to make a telephone call at the Shell Oil barge in the harbor. "Thus Endeth the first Lesson," he wrote.

At the beginning of the 1936 fishing season, the experienced fisherman "roared through the Yaculta Rapids," and "heard the most beautiful chorus of birds in the evening." He started reading Thomas Hardy's *Return of the Native*, but he decided that H. G. Wells's *Outline of History* was "more to the point." He sold a "No.1 mild-cure for eleven cents a pound, and Coho for six," and reported that his friend Edgar Lansdowne "nearly sank his boat" with fifty dollars' worth of fish. HERE ENDETH THE SECOND LESSON, he wrote on September 24, 1936, in Vancouver.

In 1937, the entries began acquiring the brief style of the more recent logbooks. He made occasional notations of something he wanted to remember, which he also did later. One that particularly interested me: "A fisherman hauls in his fish, and as each Coho comes up to the side of the boat he says to it, 'Well, I'm sorry, but you must die that I may live.' "

At the back of the book were several pages of precise diagrams of lineups in favorite fishing areas, including Goose Island, Bella Bella, and "Poverty Flats," Terry O'Connor's fishing grounds along the Vancouver Island shore west of Bull Harbour. Pasted on a back page, there was also a typed list of items needed to expedite shore time in port: food, household items, fishing and engine equipment.

I finished the logbook and started to put it away, and I dropped it. When I picked it up, I noticed a yellowed newspaper clipping pasted at the very inside of a page that I had missed. It was a clipping of a news item, from Blackpool, England. It read:

> A gold locket tumbled to a fish counter here the other day when a big halibut was being cut up. Inside the locket was found the picture of a girl in her 'teens with long flowing hair in pre-war fashion. The halibut had been caught 300 miles off the northwest coast of Scotland.

XXVIII

To get home from the Central Fishing Grounds, we had to pass through a formidable exit, the Yaculta Rapids, whose name derived from a neighboring Indian tribe. I was astonished to discover its true spelling when I looked it up, because I had never heard it pronounced any way but "Euclaetaw." We were up long before dawn to make the run, waiting for the slack tide to reverse the powerful north-going current—seven to nine knots—so we could move with that rushing water. We were at the north end, waiting to go south. There were six boats poised to go with us, four of them trollers—all with our poles up, of course. Ahead were brooding mountains, with a hovering patch of white cloud cutting across them—their dark tops appeared to be floating in space.

Everything was still. Not just the boats—the mountains, the sea, the clouds, all waited in the deep silence. The black mass of mountains with the white shroud across them and the faint orange of the sky behind their peaks created a special kind of eerie light I wanted to remember. I saw us as if we were part of a painting: the insignificance of our small boat waiting before those awesome mountains.

The sun was coming up slowly now, shining on the black water, and I looked at my watch. Six-thirty. I had asked John beforehand, so I knew that we had seven miles to go, four of them through turbulent rapids—violent swirls and eddies—ahead of us. We would be traveling with the tide at about nine knots an hour—two knots faster than our usual running speed—through a depth of twenty to forty fathoms.

We had been moving along slowly, and as the rays of the sun rose over the mountains our boat began to rock from side to side in the glistening blue-black water. We started cautiously around a point, and the rocking became more violent; I could hear the mugs clanking against one another behind me, and the teakettle sliding back and forth on the stove. I was conscious in this rocking of all our neighbors; I could see them rocking—companions in this mysterious voyage.

I looked at my watch again. Seven-thirty. John was guiding our boat through small swirls in which we twisted and rolled. Then we went around a bend and were caught up in the whirling flood of the Yacultas. Once caught, we were swept along, reeling and turning in the waters boiling up around us. Around the bend, the mountains were no longer ahead but looming on either side as we swiveled and swayed. All the boats were moving forward, moving through the rapids, rolling back and forth, twisting every which way. The powerful current sucked at the bottom of the *MoreKelp*, pulling it from side to side, wanting to turn it completely around. It was a continuing battle to stay on course.

Once, John handed the wheel to me. It moved as if it were being tugged firmly from below. I felt the driving force of the current gripping our rudder. The *MoreKelp* was responding to the tide's commands rather than to the movements of the steering wheel. The boat was going one way, but the bow seemed to want to go in another direction, as if they were separate entities. I was glad to relinquish the wheel when John took over again. Suddenly the reeling and lurching, the twisting and turning, ceased. We were in the quiet gentle water of Calm Channel, chugging comfortably along at our usual traveling speed.

Tired by afternoon, John stopped at the mouth of Desolation Sound to nap on the deck, and we drifted. I slept, too, until we were awakened by a shout from a boat that had come alongside. It was Clarence and Mary Cook, friends from Pender Harbour. "We saw your boat drifting and thought something might be wrong," Clarence said. Then they waved and went on their way.

That evening, in the oncoming darkness, we passed the gray, misty shadow of the northbound sailboat *Cherie*, with the small wraithlike figure of John's old friend, the Englishman Gwyn Gray Hill, standing on the deck, waving at us; and then we passed the great white puffs of smoke and the bright electric lights that illuminated the sky around the pulp mill at Powell River. We anchored in a place called Thunder Bay, very close to home. We started trolling shortly after dawn and were joined by our neighbor, Sonny Reid, on the *Instigator*. Neither boat caught a fish. A little more than an hour later, we were turning in among the familiar islands that mark the entrance to Pender Harbour, and were home.

XXIX

We took a shower in the house and were off again, heading for Vancouver. Our hold was filled with the salmon John had caught on the way down. I steered while John nervously made lists of people to notify when we arrived, referring to his green-leather address book for their telephone numbers. As soon as we had tied up at the Campbell Street dock, I was sent off to the public telephone booth to call them.

The whole next day, John stayed on the dock, handing out beautiful sleek salmon; visiting with friends, many of them widows of old fishing pals; and thoroughly enjoying himself, in what I found out in subsequent years was an annual ritual. By the end of the day, all the fish we had caught on the last trip were given away, and, with a sigh of relief, John said, "Now we can go and visit Geoff and Margaret Andrew!"

We took a taxi to the university section of Vancouver, stopping at a neat white house in a pleasant neighborhood. We were met at the door by a handsome gentleman with a shock of snow-white hair, and behind him, at the door, his wife, a comfortably smiling woman in a red tweed suit, with hair drawn tightly back in a knot from an intelligent, lovely face. She had just arrived home from a meeting of the Vancouver School Board, over which she had presided as president.

We joined the Andrews for a late, substantial, and delicious dinner, prepared by Margaret. We ate at a long table, in a room whose casement windows held shelves of Inuit stone sculpture that they had collected. I was in a state of shock as I sat drinking wine out of a delicate green-crystal glass, eating raspberry mousse from exquisite, antique pink china, with a fire crackling in the grate in the living room beyond. This was the familiar life I had lived before I met John. What had once been familiar was now strange—even exotic. I had gone on a long journey, from which I could never quite come back, although the Andrew house eventually became a haven, and those two wonderful people eventually became my extended family.

After dinner, John was engrossed in conversation with the teenage youngest Andrew daughter, Kate, whose education he was following; he frequently sent her books that he thought might interest her. Geoff asked me if I would like to hear how he and John met. It was a story he obviously liked to tell, and I said that of course I would.

"I'm retired now, but I was then professor of English and executive assistant to the president of the University of British Columbia, which meant that I acted in his place when he was away," Geoff began. "Once a year, in the winter, we had a course for commercial fishermen. The idea was to make it possible for anybody to improve his knowledge of commercial fishing, especially someone who had great practical knowledge but not much formal education—a beautiful example of what you can do through the extension program of a university that has an oceanography department. I went over to the extension department to open the course, and I remember being struck at once by this tall, fine-looking guy who was attending. That was John. He had this cultivated voice and distinguished bearing, and afterwards I went up and talked to him. He indicated he had a couple of sons, and was divorced, and"—with a side glance at me—"he said in a self-deprecatory way that he didn't think any woman would want to live with him, because he wanted to go his own way and wasn't very accommodating."

I murmured that I didn't find that to be so.

"He told me he had a son almost ready for university, and another coming on, and he wanted to know the cost of registering, and of board and tuition. I said, 'When you get free time in this course, come to my office, and I'll give you the information.' He did come, and at that meeting I formed the basis of a friendly relationship. I told him that when his son entered university we would like to have him come around for a meal. When Dick attended U.B.C., John came down from time to time, so we'd see him. We also saw Dick, who was in the same class with my daughter Shossie. John was interested in what Dick was learning and very proud of him. He was also fascinated with whether he himself had missed anything that Dick was getting. By the time Sean came, I had left to become director of the Association of

Universities and Colleges of Canada, in Ottawa, but John and I never lost touch. All the time I was East—and this touched me very much—he kept writing to me. I not only cherished John's letters but I enjoyed the relationship. And I still do." He sat back, smiling. "I've always thought of John as being a genuine Canadian eccentric who is continually swimming upstream except if the wind is too strong, and then he turns and swims down: a curious mixture of English and American mythology who feels so secure in his social background—he once showed me that family tree that goes back to William the Conqueror—that it has liberated him from any need to be conventional."

XXX

Our fishing life continued within the same general pattern of that first year, but some things changed. John took off the bow poles the next season. From then on, as a consequence, we had four gurdies instead of six, and only two main poles to raise and lower. I could never pull them up, although I tried; they were too heavy for me. John found having two poles less work, and he caught pretty much the same amount of fish.

John died on February 18, 1978, at The Pas, Manitoba. We were on our way, with Dick, to visit Sean and his family in northern Manitoba, where Sean was working as a geologist in a mine. Dick came west from Toronto to meet us in Winnipeg, and we stopped overnight farther north to attend the annual Trappers' Festival at The Pas, which John had been longing to do for years. That evening, we went to a dance on the Indian reserve. John immediately set off in search of the chief, and made an appointment with him for the next morning to talk about British Columbia salmonid-enhancement and future fish-conservation projects on that northern Manitoba Indian reserve. The musicians, a violinist and guitarist, sat on a raised platform, and the room was packed with people dancing or sitting at long tables, smoking and drinking. John was a terrible dancer, but he loved to dance—especially to country music—which he did with complete abandon and total joy. We danced several dances, and when we stopped I noticed how stuffy it was in the crowded room. I was worried about the lack of oxygen in a room so filled with smoke.

I said reluctantly, because we were having such a good time, "I wonder if we should leave."

"I'm completely happy," John said. "We'll leave after this next dance."

At the end of that round, when the music stopped, he said, "I can't breathe. I'm going outside."

I turned and called to Dick, and then I ran after John. I found him lying in the snow; the minute he came out into the cold air, he had dropped dead.

Dick stayed six weeks, working in the garden and on the boat, putting the *MoreKelp* together so that we could sell her. She left the harbor on May 5, 1978, with a new owner at the helm, a young fisherman who continued to report to me on her welfare until he, in turn, sold her. The next owner paid such a high price for her during a period of inflated boat and land values in B.C. that when he could no longer meet the heavy monthly mortgage payments she landed in the bailiff's office on Vancouver Island. Jimmy Reid, the boatbuilder in Garden Bay who had worked with John to extend the length of the *MoreKelp*, went over and bought her. John would have been pleased. I sometimes catch a glimpse of her in our harbor. I'm glad that the *MoreKelp* is still fishing, keeping up her reputation as a boat that catches fish.

When I came to Garden Bay to live, John took a third of his workshop down by the water and made it into a study for me. He put up a plywood partition and painted the interior on my side marine-paint white. We often conversed through the thin walls, and when he was working down at the boat, I could lean out my study window, with its spectacular view of the bay, and talk to him on the dock. He liked to make tea on the stove on the boat and bring it up to me, and frequently he took a nap on the study couch. The silence down there after he was gone was overwhelming.

Not long after John died, an old-time Pender Harbour resident, Wilf Harper, gave four acres of his land nearby for a regional park that contained a salmon-spawning stream. It was a very unstable creek, full of fallen trees and sandbars, due to logging upstream; and torrential rains regularly caused flash floods that overran its banks. In the fall, the salmon were washed away, their carcasses strewn in the woods and fields when the water receded in the spring; fertilized eggs laid in the gravelly creek bed suffered a similar fate. John organized fishermen and loggers to work every winter to build up the banks. When Wilf gave the land to the community, he also gave it its name: the John Daly Nature Park.

When John died, I thought I would of course return to the United States to live, but it was too late. I had come too far in another direction. With the new pair of eyes that John had given me, I could not go back to what I had been before. I also could not continue alone in the fishing life. There are women who do, but I had been an observer, not a participant; it was fishing with John that I loved so much. I don't think I was meant to go away; John saw to that by leaving me the house for my lifetime. How could I leave Pender Harbour when there was so much to hold me there?

Before I came to live with him, John wrote to me from the fishing grounds: "When you attain the oneness I have with the sea and the mountains, and the B.C. coast—and you will, you will—you can face anything, and alone, if you have to. This will be my hoped-for gift to you. Neither are we parted then, nor do we die."